ASTROLOGY AS ART

ASTROLOGY AS ART
Representation and Practice

Edited by
Nicholas Campion and Jennifer Zahrt

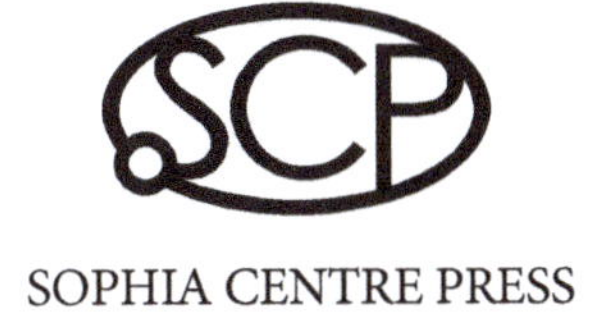

SOPHIA CENTRE PRESS

Astrology as Art
edited by Nicholas Campion and Jennifer Zahrt

© Sophia Centre Press 2018

First published in 2018.

Sophia Centre Press
University of Wales, Trinity St David
Ceredigion, Wales SA48 7ED, United Kingdom.
www.sophiacentrepress.com

Cover Illustration: Raffaello Botticcini (Attributed to) (1477–ca.1520), Detail of 'The Adoration of the Magi', ca. 1495, tempera on panel, d. 104,2 cm. Mr. and Mrs. Martin A. Ryerson Collection, 1937.997, The Art Institute of Chicago.

Publisher's Cataloging-in-Publication
(Provided by Cassidy Cataloguing Services, Inc.).

Names: Campion, Nicholas, editor. | Zahrt, Jennifer, editor.
Title: Astrology as art : representation and practice / edited by Nicholas Campion and Jennifer Zahrt.
Description: Ceredigion, Wales, United Kingdom : Sophia Centre Press, 2018. | Series: Studies in cultural astronomy and astrology ; vol. 10
Identifiers: ISBN: 978-1-907767-10-4 (paperback) | 978-1-907767-60-9 (ebook)
Subjects: LCSH: Astrology in art. | Zodiac in art.
Classification: LCC: N7745.A85 A88 2018 | DDC: 704.9491335--dc23

Printed by Lightning Source.

CONTENTS

FOREWORD

Nicholas Campion and Jennifer Zahrt

The chapters in this anthology – the latest in the Sophia Centre Press series on 'Studies in Cultural Astronomy and Astrology' – are based on papers originally presented at the 2015 Sophia Centre conference on 'Astrology as Art: Representation and Practice'. Two questions arise from the title: first 'is astrology an art?' and, second, 'how does art represent astrology?' The first question may be understood in various ways, such as 'is the act of interpreting astrological symbols an art?', or 'is the visual language used by astrologers artistic?' And then, on a deeper level, if astrological artefacts are considered to be alive, as talismans may be, then they are not representational. In that sense, how do we consider them art without misrepresenting their identity.

The definition of astrology as an art is familiar from secondary sources. Morris Jastrow in the *The Encyclopaedia Britannica* described it as 'the ancient art or science of divining the fate and future of human beings from indications given by the positions of the stars (sun, moon, and planets)'.[1] The *Concise Oxford Dictionary* defined it as '(Formerly) practical astronomy (also called *natural ~*); art of judging of reputed occult influence of stars upon human affairs (*judicial ~*)'.[2] Roger Beck wrote 'Astrology, the art of converting astronomical data (i.e., the positions of the celestial bodies) into predictions of outcomes in human affairs'.[3] And amongst astrologers, Margaret Hone, whose *The Modern Text Book of Astrology* ranked as one of the most influential textbooks of the 1950s–70s, considered art to be one way of describing astrology: 'Since there are many angles of approach to astrology as to religion or art, it is not easy to formulate a definition to suit all, but few will find disagreement with the following: *Astrology is a unique system of interpretation of the correlation of planetary action in human experience*'.[4] The key to Hone's statement, perhaps, lies in the word 'interpretation' – astrology may have an objective existence written into the fabric of the cosmos, but the astrologer has to consider, reflect, and provide meaning.

---

[1] Morris Jastrow, 'Astrology' in *The Encyclopaedia Britannica*, 10th edition (Cambridge: Cambridge University Press, 1910), p. 795.

[2] 'Astrology', *Concise Oxford Dictionary* (Oxford: Clarendon Press, 1952).

[3] 'Astrology' in *The Oxford Classical Dictionary*, 3rd edition, Simon Hornblower and Anthony Spawforth, eds., (Oxford: Oxford University Press, 1996), p 195.

[4] Margaret Hone, *The Modern Text Book of Astrology* (Romford: Fowler, 1951), p. 16.

Astrology and art are words with different and contested meanings. At its broadest we can define astrology as the observation of relationships between the sky, stars, planets, and Earth, a statement which includes both the traditional medieval branches of the discipline: natural astrology which deals with physical and seasonal cycles, and judicial astrology, in which astrologers interpret matters of human interest, offering advice, managing the present and, sometimes, predicting likely futures. Both, of course, may overlap. As Patrick Curry succinctly put it, 'In practice the line between natural and judicial astrology was constantly blurred'.[5]

Art, meanwhile, is, like judicial astrology (any astrology in which the astrologer's judgement and interpretative skills are central), is largely a human creation. It is an artefact. In the first century BCE, Cicero defined one of his two forms of divination as 'artificial', meaning it relies on a structure, a framework which the diviner creates in order to reach a logical answer to a question.[6] Popular modern examples include tarot cards and the *I Ching*. The astrological horoscope, the complex schematic diagram used by astrologers, can be another, although not necessarily so. Cicero's other form of divination was 'natural', by which he meant that it is unrestrained by reason and logic. It is spontaneous and involves a direct interaction between the diviner and the divined. Divination from dreams, trance, and forms of enchantment are examples of natural divination.

Art relates to the modern English words artifice and artificial: a work of art is artificial. It is something deliberately created and, for most of western history, has required a high level of craft. A skill has to be learnt, and if it is implemented well then the viewer, listener, or reader is moved and inspired. The practitioner of artificial divination has to aspire to such heights of they are to be successful in providing effective advice. However the history of western art since the late eighteenth century has seen first the Romantic notion that the artist must express the soul – either their own soul or the soul of nature, or the world soul. And as the twentieth century progressed, the idea began to take hold amongst some that skill and craft were irrelevant compared to the artist's all-important self-expression. We are reminded both of Cicero's natural divination in which the diviner naturally speaks the truth on the basis of a deep connection with the world, and of modern astrologers who regard themselves as speaking the truth on the basis of intuitions – ideas, thought and feelings emerging into consciousness from the psychic depths – rather than the application of the technical rules of astrology. Some artists, though, have reached a position

---

[5] Patrick Curry, *Prophecy and Power: Astrology in Early Modern England* (Oxford: Polity Press, 1989), p. 10.

[6] Cicero, *De Divinatione*, trans. W. A. Falconer (Cambridge, MA: Harvard University Press, 1929), I:lvi.

in which art has no essential existence or quality. The value of what they create is judged entirely by the audience. Astrology tips into this domain even though few dare to admit it. After all Alexander Ruperti, one of the leading astrologers of the western world in the 1970s–90s declared that

> there is not one Astrology with a capital A. In each epoch, the astrology of the time was a reflection of the kind of order each culture saw in celestial motions, of the kind of relationship the culture formulated between heaven and earth.[7]

Within mainstream western astrological thought itself runs the notion that judicial astrology has no absolute truth but evolves with culture. A medieval horoscope may resemble a Gothic cathedral, a modern one Le Corbusier. Some forms of art have moved from the representational to the conceptual. Astrological iconography conforms to both. There are artistic representations of zodiac signs and planets that may have no specific astrological use. Then there are the visual forms used by astrologers: glyphs to represent planets and drawings of horoscopes. Glyphs themselves carry concepts and some modern astrologers regard the reading of the symbol as the key act of the astrologer, as if the symbol has no fixed meaning.[8] We then move into the realm of contemporary conceptual art, in which what is or is not a work of art is as dependent on the viewer as the creator.

The boundary between representation and practice is fluid and each of the nine chapters in this volume passes between them. They explore the meanings of art and astrology, the iconography of astrology and the nature of its practice, the use of zodiac signs, and the portrayal stars and planets in literature and the visual arts. The volume opens with Martin Gansten's exploration of what exactly we mean by 'art'. Micah Ross and Suzanne Nolan then consider astrological iconography in Mesopotamian and Mediterranean culture, and the Mesoamerican worlds, respectively. John Meeks moves into literary territory with his analysis of astrological symbolism in Wolfram von Eschenbach's thirteenth-century epic, *Parzifâl*. Claudia Rousseau, Ruth Clydesdale, Spike Bucklow, and Richard Dunn then take four perspectives on the European Renaissance, a highpoint for the use of astrology through imagery,

---

[7] Alexander Ruperti, 'Dane Rudhyar – March 23, 1895–September 13, 1985 – A Seed-Man for the New Era', accessnewage.com, originally printed in *Astrological Journal* 33, no. 2 (Spring 1986): p. 55. See also http://accessnewage.com/articles/astro/rud8.htm [accessed 25 June 2017].

[8] Liz Greene, 'Signs, Signatures, and Symbols: the Languages of Heaven', in Nicholas Campion and Liz Greene eds., *Astrologies: Plurality and Diversity* (Lampeter: Sophia Centre Press, 2011), pp. 17–45; Nicholas Campion, 'Is Astrology a Symbolic Language?' in Nicholas Campion and Liz Greene, eds., *Sky and Symbol* (Lampeter: Sophia Centre Press, 2013), pp. 9–46.

metaphor, and symbol. Liesbeth Grotenhuis closes the volume by moving the debate to the modern era with her study of the symbolist painter Fernand Khnopff.

Nicholas Campion,
Associate Professor of Cosmology in Culture,
Principal Lecturer, Faculty of Humanities and the Performing Arts,
University of Wales Trinity Saint David.

Jennifer Zahrt,
Honorary Research Fellow,
Faculty of Humanities and the Performing Arts,
University of Wales Trinity Saint David.

INTRODUCTION:
ARS, TECHNĒ, ŚĀSTRA, ʿILM: WHAT'S IN A NAME?

Martin Gansten

ABSTRACT: The designation of astrology as an art raises some simple but fundamentally important historical questions. How old is this designation, and, given that earlier generations of astrologers wrote in different languages, what words did they use that we translate as 'art'? What are the connotations of those different words in their historical contexts? What other designations than 'art' would have been possible, and were such alternative labels used as well? Was astrology ever practised in cultural contexts where distinctions like that between 'art' and 'science' were not made? And if calling astrology an 'art' was not originally a question of aesthetics, was there still an aesthetic dimension to the practice of astrology, and how might we define or understand that dimension?

Introduction

We all know that language is a changeable thing, and that words may acquire new meanings over time. What I want to discuss in this chapter is what ancient, medieval and early modern astrologers are likely to have meant when they described astrology by the word *art*, or by words that we translate as 'art'. I should perhaps state at the outset that this endeavour is not an instance of what is sometimes called the etymological fallacy: I am not proposing to find out the true meaning, or even the current meaning, of either 'art' or 'astrology' by means of linguistic analysis. Rather, what I am aiming at is what every historian or philologist tries to do, namely, to read texts – in this case, astrological texts – carefully and in a culturally and contextually sensitive way.

If we look up *art* in a contemporary English dictionary, we find meanings such as these:

The expression or application of human creative skill and imagination, typically in a visual form such as painting or sculpture, producing works to be appreciated primarily for their beauty or emotional power

(**the arts**) The various branches of creative activity, such as painting, music, literature, and dance

(**arts**) Subjects of study primarily concerned with human creativity and social life, such as languages, literature, and history (as contrasted with scientific or technical subjects)

A skill at doing a specified thing, typically one acquired through practice.[1]

These consecutive levels of meaning take us not only from the more common or widespread usages to the less common, but also from contemporary to historically earlier strata. The word *art* as it is used today, even by people with radically different views about what constitutes 'good' or 'real' art, is largely to do with creativity, aesthetics, and emotion. In the older sense, however, an art is simply a craft or an acquired skill, often contrasted with whatever is natural or inborn. This is not, of course, to say that premodern societies or civilizations were strangers to expressions of creativity, aesthetics, and emotion; merely that they did not use the word *art* to refer to them – or not exclusively to them: insofar as aesthetic activity involved a skill or craft to be mastered, it could be called an art, but the word *art* did not necessarily connote aesthetics.

To be precise, most earlier societies did not use the word *art* at all, as their cultures were expressed in other languages. It may seem rather trivial to point out that, for instance, Hellenistic authors did not write in English; but when we read works in translation it is very easy to forget that every occurrence of a term like 'art' is the result of a choice made – whether consciously or unconsciously – by the translator, and that, quite often, different but equally justifiable choices could have been made. Nor do we typically find one-to-one relationships between the terms used in the original and those of the translation, at least not in the more readable translations. To illustrate, let us look at a few well-known astrological texts.

Hellenistic origins

Horoscopic astrology arose in the Hellenistic world, building on Babylonian foundations, so a good starting point would be looking at Greek words that were used to describe astrology. The Greek term most often translated as 'art' is *technē*, which has precisely the range of meanings just mentioned: '*art, skill, cunning of hand* [...] *craft*, i.e., *a set of rules, system* or *method of making* or *doing*, whether of the useful arts, or of the fine arts'.[2] This is such a broad definition that we may almost be tempted to ask: what else could they have called astrology, if not an art? And in the most influential of all Greek astrological texts, the *Tetrabiblos*, Ptolemy does indeed

---

[1] Excerpts taken from the online Oxford Dictionaries: http://www.oxforddictionaries.com/ [accessed 10 October 2018].

[2] Excerpts taken from Henry George Liddell and Robert Scott, *A Greek-English Lexicon* (Oxford: Clarendon Press, 1940).

use the designation *technē* twice, with Robbins's well-known translation rendering it as 'art'.[3] But Ptolemy also calls astrology a theory, *theōria*, which Robbins likewise translates as 'art', and *epistēmē*, which is translated as 'science'.[4] So there is some variation in terminology, more so in the original than in the translation; but nowhere does Ptolemy associate astrology with aesthetics or the fine arts. (He mentions those a few times in other contexts, calling them 'that which relates to the Muses' – which of course is the origin of our word *music*.)

However, if we turn for comparison to Ptolemy's contemporary and fellow Alexandrian, Vettius Valens, we find that he does not call astrology *technē* at all: the word occurs twenty-odd times in his *Anthologies*, but never as a designation of astrology.[5] Nonetheless, the widely circulated translation by Mark Riley – which, I should add, is by his own admission a preliminary one – uses the word 'art' as a designation of astrology quite often, to render a number of Greek terms that to me, at least, have a more theoretical ring to them.[6] The most frequent of these is *theōria* itself, which occurs ten times – twice as often as the next in order, *epistēmē*, which I (like Robbins) prefer to translate as 'science'.[7] The last three terms used by Valens, with at least one occurrence each, are *suntaxis*, *mathēma* and *doxa*, which mean something like 'system', 'learning' and 'opinion', respectively.[8] So to Valens – who, intriguingly, is often seen as a more hands-on, practical astrologer than Ptolemy – astrology is a theory, a science, a system, a form of learning, and an opinion, all of which sounds rather more abstract, even academic, than the 'art' with which Riley has somewhat loosely translated all these terms into English.

Astrology, then, is sometimes called an 'art' or 'craft' in Greek, if not as often as

---

[3] Claudius Ptolemy, *Tetrabiblos*, trans. F.E. Robbins (Cambridge, MA: Harvard University Press, 1940). A more recent critical edition of the Greek text is found in Wolfgang Hübner (ed.), *Claudii Ptolemaei opera quae exstant omnia*, vol. III.1: Ἀποτελεσματικα (Leipzig: B.G. Teubner, 1998). The use of *technē* to designate astrology is found in I 2,13 (trans. Robbins, p. 13) and I 3,18 (Robbins, p. 31); it is further implied in III 1,1 (Robbins, p. 221) by the adjective *prognōstikē* [sc. *technē*].

[4] Both in Ptolemy, *Tetrabiblos*. I 2,12 (trans. Robbins, p. 13).

[5] For the Greek text see David Pingree (ed.), *Vettii Valentis Antiocheni anthologiarum libri novem* (Leipzig: B.G. Teubner, 1986).

[6] Vettius Valens, *Anthologiae*, trans. Mark Riley, published electronically (http://www.csus.edu/indiv/r/rileymt/vettius%20valens%20entire.pdf) [accessed 10 October 2018].

[7] When the two words occur in close proximity, Riley sometimes does translate one as 'art', the other as 'science'. *Theōria* as a designation of astrology occurs at Vett. *Anth.* III 9,5; V 2,10; V 8,110; VI 8,9; IX 1,3; IX 3,1; IX 12,2; IX 15,1; IX 15,11; IX 16,6. *Epistēmē* occurs at II 36,19; V 8,110; VIII 5,20; IX 9,10; IX 12,3.

[8] *Suntaxis* as a designation of astrology occurs at Vett. *Anth.* IV 11,11; *mathēma*, at V 6,22; and *doxa*, at VIII 5,20.

the translations might make us think; but that is only one of several designations, and neither the most frequent nor the most specific one. Now, if we travel from Alexandria across the Mediterranean to Sicily, we find, a couple of centuries later, the Roman lawyer Julius Firmicus Maternus writing on Hellenistic astrology in Latin.[9] The English word *art* is, of course, derived from the Latin *ars*, and Firmicus does call astrology *ars* a few times.[10] This Latin term covers more or less the same semantic range as the Greek *technē*, with 'acquired skill' or 'craft' being among the most frequent meanings – again, there is no necessary creative or aesthetic dimension here. Much more often, though, Firmicus calls astrology *scientia*, which is the origin of our word *science*.[11] Just as *ars* does not correspond very closely in meaning to the modern concept of art, so *scientia* doesn't mean 'science' in the modern sense; but it does mean 'knowledge': it is a matter of understanding something, rather than of creating or enjoying it. Another word that Firmicus is fond of using to describe astrology – so fond, in fact, that it is the title of his book – is the Greek loanword *mathēsis*. Like the *mathēma* used by Valens, it means 'learning' and is related to our word *mathematics*, and of course to *mathematicus*, which for many centuries was one of the most common Latin words for 'astrologer'.

Perhaps more surprising to many is the fact that Firmicus occasionally uses the word *religio* as a designation of astrology.[12] Again, this does not mean that he viewed astrology as 'a religion' in the modern sense of a more or less exclusive system of observances and beliefs; but it does connect astrology with what we would call religious attitudes and experiences, as does the adjective *divina*, 'divine', which he sometimes uses to qualify *ars* or *scientia*.[13] Perhaps this is where astrology in the classical world most closely approximates 'art' in our sense of aesthetic creativity and enjoyment – but let us return to that point in a moment.

India, the Islamic world, and Europe

By the time Firmicus wrote his *Mathesis*, horoscopic astrology had most likely already made its way from the Greek-speaking world to India, where it soon took root.

---

[9] For the Latin text see W. Kroll and F. Skutsch (ed.), *Iulii Firmici Materni Matheseos libri VIII*, 2 vols. (Leipzig: B.G. Teubner, 1897–1913; repr. 1968).

[10] Firm. *Math.* I 2,5; II 30,10; IV prooem.,5; VII 26,12; VIII 5,1; VIII 25,10.

[11] Firm. *Math.* I 1,1–2; I 3,2–3; I 3,5; I 4,1; I 4,5; I 4,14; I 5,1; II praef.,1; II 20,14; II 30,1; II 30,14–15; IV prooem.,5; IV 1,10; IV 22,1; V praef.,1; VIII 5,3; VIII 33,2

[12] Firm. *Math.* II 30,14; VIII 5,1.

[13] Firm. *Math.* I 1,1; I 3,3; I 3,5; II 30,1–2; II 30,10; II 30,14; IV prooem.,5; IV 22,1; V praef.,1; VII 26,12.

The Sanskrit culture of classical India has no concept precisely analogous to *technē* or *ars*, but it does have a category somewhat closer to our modern idea of art: *kalā*, which covers a number of what we would call the fine arts, including singing, music, dance, drama, painting and sculpture – although it also covers a number of practical skills, mostly associated with pleasures and amusements, that may not strike westerners as equally fine or important, such as dressing hair; arranging flowers; putting on costume, jewellery and perfumes; preparing drinks and dainties; playing games, telling riddles and doing magic tricks; organizing cockfights, teaching parrots to talk; and so forth.[14] Some of the more complex arts can be classified in their theoretical aspects as *śāstras*, a word that denotes any systematic body of teaching (and the texts expounding it); but most *śāstras* are not arts (*kalā*).

Astrology, like medicine, is uniformly viewed as a *śāstra* – *jyotiḥśāstra*, 'the science of lights' – or a *vidyā*, field of learning; but never, despite its practical applicability, as a *kalā* or art. In India, then, we have an ancient culture that does possess a separate category corresponding more closely to 'art' in our modern sense, but which unambiguously excludes astrology from that category. Nevertheless this is arguably the culture where horoscopic astrology has thrived and developed for longer, and achieved more widespread acceptance, than anywhere else in the world.

Greek and Sanskrit astrological traditions, along with Syriac and, above all, Persian ones, formed the basis of the great Arabic astrological synthesis that began in the eighth century. I am no Arabist and cannot claim the same familiarity with Arabic-language astrological works as with their Greek and Indian counterparts, but I find that just the titles of these works tell us a great deal.[15] Phrases like *ʿilm an-nujūm* ('the science of the stars'), *aḥkām an-nujūm* ('the judgements of the stars') and *ṣinā ʿat an-nujūm* ('the craft of the stars') recur constantly in varying combinations or, one is tempted to say, constellations.[16] Of these terms, my impression is that the least frequent is *ṣinā ʿa* – meaning 'craft' and thus corresponding most closely to the Greek *technē* or Latin *ars* – though the Arabic astrological corpus is a large one and nobody, to my knowledge, has attempted to produce a concordance to it. Certainly *ʿilm* and *aḥkām* are very frequent, as are the corresponding *scientia* and *iudicia* in the mostly twelfth-century Latin translations from the Arabic. This choice

---

[14] For a more extensive list of Indian 'arts' see, e.g., *Kāmasūtra* 1.3.15. Several editions of the Sanskrit text exist, but none of them critical in the strict sense. A passable translation is: Vatsyayana Mallanaga, *Kamasutra*, trans. Wendy Doniger and Sudhir Kakar (Oxford and New York: Oxford University Press, 2002).

[15] For an overview of the available literature, see Fuat Sezgin, *Geschichte des Arabischen Schrifttums*, Band VII (Leiden: E.J. Brill, 1979).

[16] For a wealth of examples see Sezgin, *Geschichte, passim*.

of words again emphasizes the role of astrology as an intellectual endeavour rather than an aesthetic one.

Following on the translations from the Arabic, original works on astrology began to appear in western Europe, eventually alongside direct translations from the Greek. Most of the published works of the Renaissance and early modern period were still written in Latin, but in the sixteenth and seventeenth centuries, publications in the vernaculars began to appear. By and large, these new contributions to astrological literature preserved the terminology employed by the early Latin translators and spoke, in their various languages, of the 'science' and 'judgement' and, occasionally, 'art' of astrology.

To an English-speaking readership, no doubt the most familiar of the early modern authors in the field is William Lilly. His writings are of particular interest to us, not only because they are numerous and influential, but because they reverse the general trend to which I have referred: whether as a reflection of the English idiom of his time or simply of his personal preferences, Lilly was very fond indeed of the word *art*. For an illustration, we need only look to the address to the reader at the beginning of his three-volume *Christian Astrology*: in just seven quarto pages he manages to call astrology an 'art' no less than sixteen times.[17] We may compare this to the designation 'science', which occurs only twice in the whole of the first book.

Lilly certainly helped popularize the English use of *art* as applied to astrology; but what did that term mean to him? It was during his lifetime in the seventeenth century that the word began to acquire something of its modern sense, relating particularly to painting, sculpture, and the like. However, a close reading of Lilly's work fails to turn up any instances of this aestheticized usage: when he speaks of *art*, it is always in the older, broader sense of the word, as a skill or craft. An illuminating instance is his discussion on finding a person's occupation from the horoscope, in which *profession*, *magistery*, *trade*, and *art* are all used synonymously.[18] An art, to Lilly, is simply whatever one is trained to do, including astrology; and an artist – a word he sometimes uses to refer to astrologers – is anyone skilled in his own art.

Inspiration and the aesthetics of astonishment

To sum up the views of Hellenistic civilization, classical India, the Islamic Golden Age and pre-industrial Europe in a few words, then, astrology can be called an art in the sense of a craft to be mastered, but hardly in the sense of a 'fine' art or creative

---

[17] William Lilly, *Christian Astrology* (London: Tho. Budenell, 1647), vol. 1, pp. 5–12.
[18] Lilly, *Christian Astrology*, vol. 3, pp. 624–34.

self-expression; and the majority of terms used to describe astrology across these cultures are in fact what in today's popular idiom might be termed left-brain ones: knowledge, system, science, learning, opinion, judgement, and so forth.

But let us pause to consider the word I just introduced into that last sentence: *self-expression*. If the aestheticized use of the term *art* is modern, the widespread understanding of artistic creativity as an expression of one's individual personality or unique vision is, historically speaking, very recent. A great deal has already been written about art in this sense being a product of European late modernity, and I shall not add to that discussion here. The point I do want to address, just briefly, is how creative and aesthetic experience can transcend the individual perspective in a way that may also be relevant to the practice of astrology.

I deliberately say 'creative and aesthetic' so as to cover both the act of creating or performing a work of art (in the modern sense) and of actively appreciating or participating in it. We are familiar, of course, with the concept of inspiration, although the word is typically used in a watered-down sense today. The exception is in religious discourse, where for a text to be inspired still means that it was dictated by or produced under the direct influence of a deity, who 'breathed into' the writer – the literal meaning of *inspiration*. Artistic inspiration, too, was traditionally viewed in this light, as the manifestation of a power beyond the artist himself, often represented by the semi-divine Muses. Whatever one's religious beliefs, this sense of transcending the everyday self and temporarily gaining access to a trans- or suprapersonal realm is still experienced by some creative artists.

But such transporting experiences are not necessarily limited to the creators of art: in fact, some aesthetic philosophers claim them as the domain of the connoisseur rather than of the artist. Such was the view of Abhinavagupta, who lived about a millennium ago and who is still one of the most celebrated Indian thinkers on aesthetics.[19] Aesthetic experience is known in Sanskrit as *rasa*, 'flavour', and according to Abhinavagupta, its essence is 'astonishment', *camatkāra* (a term variously rendered by translators as 'rapture', 'wonder', or even 'epiphany'). Significantly, he uses the same word to describe the experience of spiritual illumination, when personal consciousness expands in a shock of realization or, more accurately, recognition of ultimate reality. Although the two forms of experience – mystical and aesthetic – are not identical, Abhinavagupta considers them to be analogous; and it is precisely the universalization of emotion in art, liberating it from the confines of individuality, that makes such lofty raptures possible.

---

[19] For an introduction to Abhinavagupta's aesthetics based on the most recent scholarship see Sheldon Pollock, *A Rasa Reader* (New York: Columbia University Press, 2016), pp. 181–238.

Now, it may be argued – and in fact I would argue – that rapturous epiphanies are not confined to the realm of what we typically consider to be aesthetics, firmly grounded in visual, auditory or other sensory experience: they can also occur in contexts that are preeminently rational, such as philosophy or mathematics. Indeed, it is not uncommon for discoveries in these fields to be described in aesthetic terms by those who understand them, as 'elegant' or even 'beautiful'. If there is an aesthetic dimension to the traditional practice of astrology – and I think there is – then I believe it is more akin to delight in the beauty of a mathematical proof than to the enjoyment of a fine painting or piece of music. And this shift from the sensory to the intellectual realm makes the dividing line between aesthetic and religious experience a very thin one. Perhaps it is not even a meaningful distinction.

The ways in which astrologers express their experiences of awe and delight naturally vary with religious and cultural contexts. In the Hellenistic period, one of the most vocal authors was Firmicus Maternus, mentioned above, who called astrology a *religio* and *divina scientia*. Near the beginning of his work, he writes:

> To whom does the whole nature of divinity both show and produce itself if not to the *mind*, which, having come forth from the celestial fire, has directed itself toward the ruling and direction of earthly fragility?
>
> This one handed down the plan of this science. This one devised the computations. This one demonstrated the courses, regressions, stations, associations, augments, risings, and settings of the Sun and the Moon and the other stars [...] And constituted in the fragility of an earthly body, he did not forget, but remembered to hand down all these things with the awareness of his own brief majesty. [...]
>
> For the stars have their own sense and divine prudence, because, animated by the pure concept of divinity, they conform with unceasing agreement to [the will of] that highest and master God, who arranged all things with a perpetual disposition of law, in order to preserve the order of his creation everlastingly.[20]

Firmicus' vision is one of ordered beauty, a *kosmos* in the full sense of the word, reflected kaleidoscopically for the astrologer in an infinite number of individual horoscopes. And this, I believe, is the point at which the traditional art of astrology can become both a mystical and, in the deepest sense, an aesthetic experience.

---

[20] Julius Firmicus Maternus, *Mathesis*, trans. James Herschel Holden (Tempe, AZ: American Federation of Astrologers, 2011), pp. 15–16, 19–20.

THE ICONOGRAPHY OF LIBRA AND PISCES IN MESOPOTAMIA, EGYPT AND GREECE

Micah Ross

ABSTRACT: The division of the sky into twelve sections began in Mesopotamia and establishes a continuity among the astrological cultures of Mesopotamia, Egypt, Greece, and India. The iconography and names of the individual signs can vary, but they also preserve thematic unity. For instance, names of constellations often refer to the object imagined among the stars, and in the case of the zodiac, the iconography and the linguistic representation often overlap. Two examples illustrate this variation and continuity. The iconography of Pisces details a path from local variants to standardization, while the iconography and terminology of Libra raises unresolved questions. When they entered foreign cultures, the names were subject to linguistic borrowing, and the iconography was prone to cultural assimilation. While the iconography of these two signs seems only tangentially related, Mesopotamians referred to these signs by similar words. Mesopotamians referred to the zodiacal signs by different names in different genres. Thus, the word '*zibbāti*' can indicate Pisces, and the word '*zibānītu*' can refer to Libra. The derivation of these names from the same root serves as a point of departure for a consideration of the representation of these signs, as Egypt and Greece adapted the names and images of Libra and Pisces in interesting ways.

Preliminaries

The standard hypothesis postulates that the zodiac entered Greece in the fifth century but arrived in Egypt in the third or fourth century. A new hypothesis of zodiacal transmission which proposes that the zodiac entered both Egypt and Greece at roughly the same time necessitates a reassessment of the development of zodiacal iconography.[1] The hypothesis of a simultaneous introduction occasions a comparison of Egyptian iconography with Mesopotamian sources. To this end, the Sophia Centre has fostered an analysis of Gemini.[2] The programme in Cultural Astronomy and Astrology at the University of Wales Trinity Saint David has also hosted seminars on Aries and Leo. With the analysis of Libra and Pisces, the Sophia Centre has supported a reconsideration of the iconography of nearly half the zodiac. These investigations do not directly compare the elements of the constellations with the stars plotted in uranographies – the maps of the night sky which identify particular stars.

---

[1] Micah Ross, 'The Role of Alexander in the Transmission of the Zodiac', in Volker Grieb, Krzysztof Nawotka and Agnieszka Wojciechowska, eds., *Alexander the Great and Egypt: History, Art, Tradition*, Philippika 74 (Wiesbaden: Harrassowitz, 2014), pp. 287–307.

[2] Micah Ross, 'A Study in the Early Iconography of Gemini', in Nicholas Campion and Dorian Gieseler Greenbaum, eds., *Astrology in Time and Place* (Newcastle-upon-Tyne: Cambridge Scholars Press, 2015), pp. 95–108.

Rather, they examine zodiacal iconography, including the names of zodiacal signs and their identification with culturally familiar elements such as stock characters, gods, and heroes.

Pisces

Mesopotamia

Mesopotamian astronomers developed the zodiacal coordinate system by standardizing their proto-zodiacal constellations. In the case of the region of the sky destined to become Pisces, these astronomers combined pre-existing elements, among which no compositional unity need be presumed. At different times, astronomers identified these precursors to Pisces with slightly different astronomical objects or observational dimensions, but they preserved iconographic elements through the star names used in several astronomical genres.

In the Middle Assyrian 'Astrolabes' (or, '*Three Stars Each*' texts), Mesopotamian astronomers recorded four elements which would later compose the zodiacal sign of Pisces.[3] These texts associate the descriptive names of three stars, asterisms, or celestial objects with each month of the year. As precursors to the twelvefold division of the zodiac, they assign one name to each of the northern, middle, and southern portions of the horizon. An early example of this genre, now called 'Astrolabe B', lists in four consecutive months four elements which later coalesced into Pisces.[4] 'Astrolabe B' assigns the middle section of the horizon in Month XI to 'the Swallow (SIM.MAḪ [=*šinūnūtu*])', associates the southern portion of the horizon in Month XII with 'the Fish (KU$_6$ [=*nūnu*])', identifies the southern part of the horizon during Month I with 'the Field (AŠ.GÁN [=*iku*])', and connects the northern band of the horizon in Month II with a star named after the goddess *Anunītum*. Although these elements would later be assimilated as a single constellation associated with one month, this precursor to the zodiac named four asterisms distributed over four months. These associations of names, months and sections of the horizon passed unaltered into the astrological compendium titled MUL.APIN, composed around 1000 BCE but in circulation until at least 687 BCE.[5]

---

[3] Wayne Horowitz, *The Three Stars Each: The Astrolabes and Related Texts*, Archiv für Orienforschung Beiheft 33 (Vienna: Institut für Orientalistik der Universität Wien, 2014).

[4] Erica Reiner and David Pingree, *Enūma Anu Enlil: Tablets 50—51*, Babylonian Planetary Omens 2; Bibliotheca Mesopotamica 2.2 (Malibu: Undena, 1981), 4.

[5] Hermann Hunger and David Pingree, *MUL.APIN: An Astronomical Compendium in Cuneiform*, Archiv für Orientforschung Beiheft 24 (Vienna: Institut für Orientalistik der Universität Wien, 1989).

Also in the seventh century, Mesopotamian astronomers began to collect nightly astronomical observations in a genre of texts which they called '*naṣāru ša ginê* (Regular Observations)' but which modern scholars have titled *Astronomical Diaries*. Early diaries described astronomical observations in terms of 'predictable stars' (MUL ŠID.MEŠ [=*kakkabū minâti*]), now called 'Normal Stars'.[6] Because stars are distributed unevenly with unequal magnitudes, several gaps occurred among these Normal Stars. Most of Pisces falls in a dark and unremarkable portion of the night sky, but the name of the sole proto-zodiacal Normal Star from the region of Pisces preserves another iconographic element: 'the bright Star of the Tether of the Fish (MÚL[=*kakkabū*] KUR[=*napaḫu*] šá DUR[=*rikis*] *nu-nu*)'. The Normal Stars add a fifth element to the iconographic elements of the '*Three Stars Each*' texts.

In 652 BCE, the earliest *Astronomical Diary* referred to a constellation in the region of Pisces as [mul]KUN.ME[=*zibbāti*], the Tails.[7] This early reference probably designates a constellation (which spanned 41°, according to Ptolemy).[8] At this time, the zodiacal signs as a metrological unit of 30° were still developing. The name indicates a constellation by the preceding star-determinative, but the name – even though it ends with a Sumerian plural – clarifies neither the exact number of tails nor the animals to which the tails belong. Orthographic variants include the Sumerian logograms KUN and KUN.MEŠ. After Mesopotamian astronomers had established zodiacal signs as the metrological standard of *Astronomical Diaries* by 454 BCE, the abbreviations ZIB.ME and ZIB occur, both abbreviations of the Akkadian word *zibbāti* (tails).[9] Whereas modern astronomers are familiar with the fish of Pisces, Mesopotamians knew 'the Tails', first as a constellation, later as a zodiacal sign.

Mesopotamian astronomers relied on lists and tables more than maps and images. In these texts, astronomers try to clarify iconographic attributes with names, but cuneiform compositions can be ambiguous. Modern scholars share a tacit understanding of Mesopotamian uranography in their translations. For example, MUL. APIN lists the stars by which the Moon passed. Pingree and Hunger counted 17 stars including the 'tails of the Swallow'.[10] Wayne Horowitz found 18 stars by separating

---

[6] Abraham J. Sachs and Hermann Hunger, *Diaries from 652 BC to 252 BC*, vol. 1 of *Astronomical Diaries and Related Texts from Babylonia*, Denkschriften der Philosophisch-Historische Klasse 195 (Vienna: Österreichischen Akademie der Wissenschaften, 1988) pp. 17–19.

[7] Sachs and Hunger, *Diaries*, vol. 1, p.42.

[8] Ptol. *Alm.*, 8.1.

[9] Francesca Rochberg, *Babylonian Horoscopes*, Transactions of the American Philosophical Society n.s. 88.1 (Philadelphia: American Philosophical Society, 1998), p. 29.

[10] Hunger and Pingree, *MUL.APIN*, p. 144.

this name into 'the Tails' and 'the Swallow'.[11] Likewise, in a study of the cryptic Dal-banna Text (700–400 BCE), Hunger and Pingree identified *Anunitu* as a fish and translated '*abri Anuniti*' as 'the fin of *Anunitu*'.[12] In other contexts, though, *abru* also denotes a 'wing'. Because both fish and fowl have *zibbāti* and *abri*, modern recon-structions must integrate astronomical sources and relatively rare Mesopotamian graphical depictions.

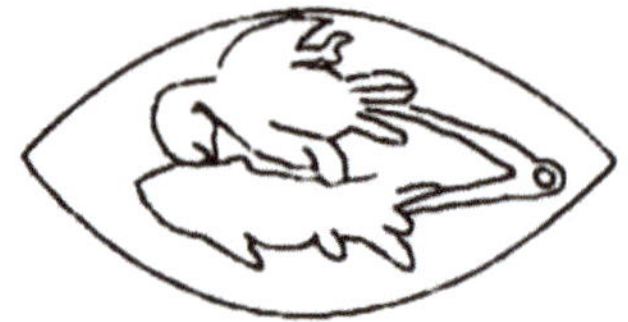

FIGURE 2.1: Impression of Pisces on a Tablet (229 BCE, Uruk)[13]

In an examination of Hellenistic cylinder seals, Ronald Wallenfels notes that the two tails of Pisces were not two fish, but a fish and a swallow.[14] This disparity in species may explain the use of a plural noun, rather than a grammatical dual, frequently used for pairs of objects such as hands or eyes. Another disparity characterizes the iconography: the bird is a swallow, but the species of fish is not clarified. Indeed, *nūnu* may refer either to freshwater fish from rivers or fish from the sea.[15] Wallen-fels further identifies a ribbon which joins the tails so that both animals point their heads in the same direction. Wallenfels suggests the composition recalls the offering strings of birds and fish. He also lists three factors which may have resulted in the change from a fish and a bird to two fish: the fact that both fish and birds have tails, the tendency to abbreviate the sign with a single fish, and 'the existence of the vo-cable *sinuntu* which, like its semantic if not etymological equivalent *šinunūtu*, not only denoted . . . a swallow, but additionally specified a flying-fish called the "swal-lowfish."[16]' Indeed, some element of wordplay may unify the elements of Pisces. The

---

[11] Wayne Horowitz, *Mesopotamian Cosmic Geography* (Winona Lake: Eisenbrauns, 1998), p. 171, n. 33.

[12] Hermann Hunger and David Pingree, *Astral Sciences in Mesopotamia*, Handbuch der Orientalis-tik. Erste Abteilung. Der Nahe und Mittlere Osten; 44 (Leiden: Brill, 1999), p. 106.

[13] MLC 2124a; Ronald Wallenfels, 'Zodiacal Signs among the Seal Impressions from Hellenistic Uruk', in Mark Cohen, Daniel Snell and David Weisberg, eds., *The Tablet and the Scroll* (Bethesda: CDL Press, 1993), p. 287, fig. 15; A.T. Clay, *Babylonian Records in the Library of J. Pierpont Morgan* vol. 2 (New Haven: Yale University Press, 1913), pl. VII, no. 207.

[14] Wallenfels, 'Signs', pp. 281–89.

[15] Ignace Gelb, *Assyrian Dictionary* (Chicago: University of Chicago Press, 1956–2006), s.v. *nūnu*, 1.a and 1.f.

[16] Wallenfels, 'Signs', p. 287.

word for fish, the species of bird, and the goddess *Anunitu* all contain *–nunu–* or *–nuni–*. Perhaps the wordplay suggests that each word contained /nun/, a closed (CVC) syllable; perhaps the assonance is accidental. Whatever motivated the development of the zodiacal sign, the Mesopotamian iconography of Pisces contained a fish and a bird.

Egypt

An exciting claim periodically surfaces in popular accounts of the zodiac: the fish of Pisces originated in Egypt.[17] This claim may be traced to Cyril Fagan, who wrote that '*Hnwy* [*sic*], the "Two fishes," beautifully and most convincingly tallies with Pisces, the 'Two Fish'.[18] Popular accounts often report the conclusion but separate the resolution from the context. Fagan proposed a reinterpretation of the decans into a system which he called 'pentades'. By this theory, the position of the Sun was marked by the rising and setting stars.

> Treated as a pentade it [=*Hnwy*] synchronizes snugly and nicely with the constellation Pisces. Khonuy (Pisces) rose at sunset during August, when the Sun was in the opposite constellation, Virgo. At this time the inundation was so excessive as to turn almost the whole of Egypt into a sea, abounding in fish and river life and shipping was the only means of transport.[19]

Although astronomically correct, Fagan has not enjoyed wide acceptance in academic circles, and the explanation is difficult to reconcile with the piecemeal development of zodiacal signs in Mesopotamia. Even though modern academics rightfully reject Egypt as the origin of the zodiacal constellation and the zodiacal sign, Fagan's argument appeals to popular audiences because Egypt spawned the modern zodiacal iconography of Pisces.

The theory by Fagan approaches circularity. For him, the correspondence of the decan *Hnwy* and the zodiacal sign of Pisces confirms the pentade system – but the correspondence would not exist without the reinterpretation. The piscine character of the decan *Hnwy* is far from certain. The name of the decan often indicates its dual number by two 'harpoon-head of bone' hieroglyphs (T19 in Gardiner's Sign List).[20]

---

[17] Cf. Ariel Guttman and Kenneth Johnson, *Mythic Astrology: Archetypal Powers in the Horoscope* (St. Paul, MN: Llewellyn, 1993), p. 357.

[18] Cyril Fagan, *Astrological Origins* (St. Paul, MN: Llewellyn, 1971), p. 57.

[19] Fagan, *Origins*, p. 115.

20 Adolf Erman and Hermann Grapow, *Wörterbuch der Aegyptischen Sprache*, vol. 3, (Berlin:

Some forms of this determinative resemble forms of hieroglyphs for fish but the 'harpoon-head of bone' relates to burials. Among the variant writings of *ḫnwy*, the Tomb of Ramses II D clarifies that the decan relates to burials by adding two 'tongue of land' (N 21) determinatives. Thus, the name of the decan may derive from the noun *ḫnw*, 'resting place'.

Arguing against an abrupt dismissal of Fagan's proposal, the Egyptian tradition of decorative motifs seemingly provides external confirmation of an early appearance of Pisces. Long before Mesopotamian astronomers standardized the zodiacal signs, Egyptian artists had developed the iconography of Pisces in which two fish are shown oriented in opposite directions, as at the bottom of the so-called 'marsh bowls'. Another explanation derives from the interplay of astrology and art: before their reception of Mesopotamian astronomy, Egyptian artists often composed images of two inverted fish according to playful geometric symmetry. After the advent of zodiacal signs from Mesopotamia, Egyptian astronomers assimilated zodiacal signs with familiar artistic motifs. In this way, Egypt hatched the zodiacal iconography of Pisces, if not the sign itself.

When zodiacal signs appeared in Egypt, the fish of Pisces constituted an invasive species. As novice students of Egyptian hieroglyphics quickly learn, Egyptians referred to fish and birds with precision. Each piscine hieroglyph specifies a different kind of fish. For example, Egyptians called the inverted fish in FIGURE 2.2 '*int*', the *Tilapia nilotica*. The distinction is important because hieroglyphs of particular fish could be used as phonetic elements in other words. When Egyptian scribes introduced zodiacal signs, they also introduced a new fish, the *tbḫ*.

The earliest Demotic text to list the zodiacal signs, Strasbourg D 521 (TM 92642), connects 'the first month of Emergence' (*tpy prt*) with the new species *tbḫ.w*.[21] Like the *tā marbūṭah* of Arabic, the Egyptian feminine suffix –t often passed unpronounced. In the case of *tbḫ*, scribes indicated the need to vocalize the terminal –t by writing '*ḫ*', derived from the pestle hieroglyph (U33).[22] The suffix –*w* indicates plurality and would have likely passed unvocalized. This word may represent an attempt to transliterate the Akkadian *zibbāti*.

Most studies on Egyptian linguistic borrowing identify the New Kingdom and the Third Intermediate Period as a period of borrowing from Semitic languages such as

---

Akademie Verlag, 1971), p. 287, s.v. ḫn; Alan Gardiner, *Egyptian Grammar*, third edition, revised (Oxford: Oxford University Press, 2007), p. 514.

[21] Otto Neugebauer, 'Demotic Horoscopes', *Journal of the American Oriental Society* 63, no. 2 (1943): pl. 3.

[22] Gardiner, *Grammar*, p. 520.

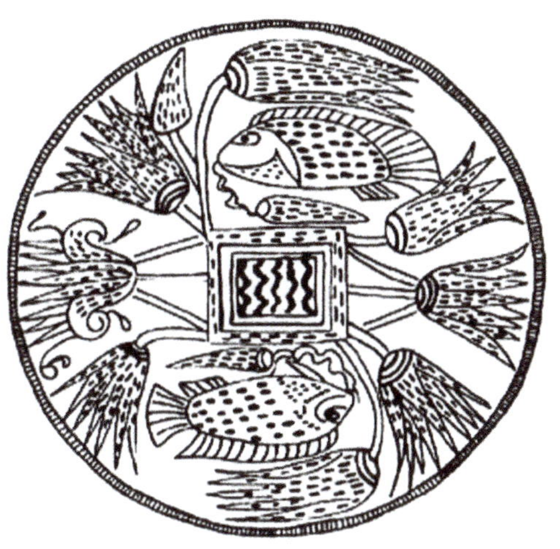

FIGURE 2.2 (left): Egyptian Motif of Inverted Fish (Dynasty 18 [1450–1400 BCE], Western Thebes). Walters Art Museum 48.400, via Wikimedia Commons.

FIGURE 2.3 (below): Motif of Inverted Fish with Pool (Dynasty 18 [c. -1550–1300]) Unknown Accession Number, Museum of Egyptian Antiquities, Cairo, Egypt; Reproduced in Eva Wilson, Ancient Egyptian Designs for Artists and Craftspeople, (Mineola, NY: Dover, 1987), p. 111, fig. 83.

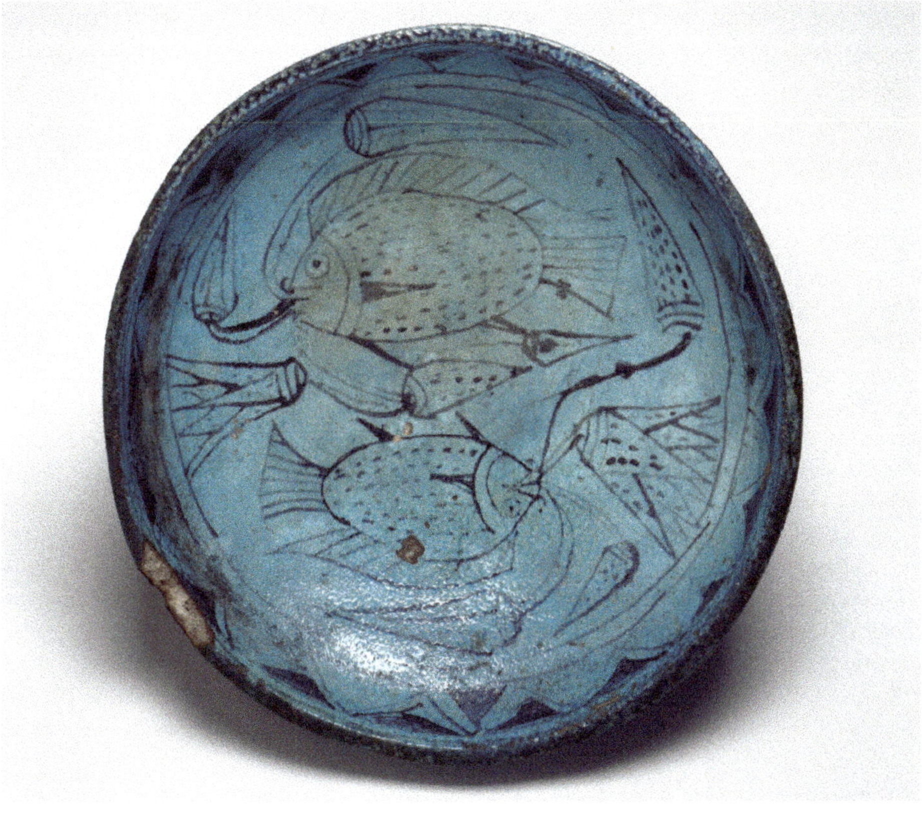

Akkadian.[23] During the Demotic phase of the Egyptian language, most of the limited linguistic borrowing was from Greek.[24] A Demotic borrowing of *tbṯ* from Akkadian subverts this trend but satisfies all of the four points indicative of a Semitic loanword, as established by Jean Winand.[25] First, the word is written phonetically. While this point carries less weight in Demotic than in Late Egyptian for which Winand composed his rules, the Egyptian scribes who referred to Pisces did not typically use an ideogram, as they did for Aries, Taurus, Gemini, Leo, Virgo, Scorpio, Sagittarius, and Aquarius. Secondly, the word should not appear in earlier Egyptian texts. As noted above, *tbṯ* cannot be equated with an indigenous fish. Third, the loanword must not have an etymological derivation within Egyptian. An attempt to derive *tbṯ* from *tpi* (an unknown fish) has been declared 'unlikely' because of gender.[26] (In short, *tpi* is a masculine word, while *tbṯ* is feminine in form but occurs with masculine pronouns. This variability prompted Dimitri Meeks to postulate a precursor **tbd* or **dbd*, which would have been rendered as /tbt/ due to Demotic sound changes.[27] Loanwords also obfuscate gender when morphological gender does not match grammatical gender.) Finally, *tbṯ* 'has Semitic cognate(s) that can be validated at the (morpho-)phonological level *and* at the semantic level.'[28] The only phonological difference is that the initial voiced, dental fricative of Akkadian (/z/) has been approximated in Egyptian by the voiceless, dental stop of Demotic (/t/), but Demotic did not differentiate the voiced and voiceless dental fricatives /z/ and /s/ and had merged the voiced and voiceless dental stops /d/ and /t/. Faced with a voiced, dental fricative (/z/), a Demotic scribe had two options: the voiceless, dental stop /t/ or the voiceless dental fricative /s/.

---

[23] Max Burchardt, *Die altkanaanäischen fremdworte und eigennamen im aegyptischen* (Leipzig: J.C. Hinrichs, 1909–1910); William Albright, *The Vocalization of the Egyptian Syllabic Orthography*, American Oriental Series 5 (New Haven: American Oriental Society, 1934); Wolfgang Helck, *Die Beziehungen Ägyptens zu Vorderasien im 3. und 2. Jahrtausend v. Chr.*, Ägyptologische Abhandlungen 5, (Wiesbaden: Harrassowitz, 1962); James Hoch, *Semitic Words in Egyptian Texts of the New Kingdom and Third Intermediate Period* (Princeton: Princeton University Press, 1994).

[24] Willy Clarysse, 'Greek Loanwords in Demotic', in Sven Vleeming, ed., *Aspects of Demotic Lexicography*, Studia Demotica 1 (Louvain: Peeters, 1987), pp. 9–33.

[25] Jean Winand, 'Identifying Semitic Loanwords in Late Egyptian', in Peter Dils, et al., eds., *Language Contact and Bilingualism in antiquity: What Linguistic Borrowing into Coptic can tell us about it*, Lingua Aegyptia, in press.

[26] Penelope Wilson, *A Ptolemaic Lexicon*, (Louvain: Peeters, 1997), p. 1134, s.v. *tbt*.

[27] Dimitri Meeks, *Le grand texte des donations au temple d'Edfoul*, Bibliothèque d'Études de l'Institut Français d'Archéologie Orientale du Caire 59 (Cairo: L'Institut Français d'Archéologie Orientale du Caire, 1972), p.90.

[28] Winand, *Loanwords*, in press.

Arguing against the interpretation of *tbṱ* as a loanword is the appearance of *tbṱ* outside of astronomical contexts.[29] Formerly, editors read *tbṱ* in the London-Leiden Magical Papyrus (TM 55955), but recent editions have reinterpreted this word as *tšr* (red), leaving only three instances of *tbṱ* being used in non-astronomical contexts.[30] First, the word appears in *s n tbṱ* (fish-monger), used in a Demotic text of 229 BCE (TM 44106).[31] This use survives in Coptic and had a Greek parallel. Secondly, the word appears in *tni tbṱ* (fish tax) in a second century receipt (TM 45622).[32] Finally, the first century outer retaining wall of the temple of Edfu contains a Donation Text which makes geographic reference to 'the well of the fish' (*tȝ gt n tȝ tbṱ*). Barring reinterpretations such as that in the London-Leiden Magical Papyrus, these instances suggest that *tbṱ* lay claim to a larger semantic range than might be expected of a borrowed technical term. On the other hand, this use of *tbṱ* may reflect an attempt to construct Demotic terms analogous to Greek words which refer to fish generally. After all, *tbṱ* had already been accepted as equivalent to ἰχθύς in astronomical contexts.

Despite the prevalence of linguistic borrowing between Greek and Demotic, the depiction of Pisces in the round zodiac of Dendera adheres closely to the Babylonian iconography. (See FIGURE 2.4.) Here, the two fish face the same direction, a cord connects the tails of the two fish, and a pool (or perhaps a well) lies between them. Although Mesopotamian astronomers maintained the Field (*Iku*) as a separate constellation, the Egyptian iconography often incorporates this once independent constellation as a rectangular body of water between the two fish.[33] The rectangular border encloses zig-zag lines, which are hieroglyphic ideograms (N 35) for water. Although the body of water conforms to the Mesopotamian placement of the constellation the Field, this element of the iconography also appears in pre-zodiacal Egyptian fish motifs. (See FIGURE 2.3.) In this way, Egyptians found an apt analogy for the Mesopotamian uranography. The rectangular zodiac of Dendera preserves the pool, albeit with no cord. (See FIGURE 2.5.) Later depictions of Pisces generally

---

[29] Janet Johnson, *The Demotic Dictionary of the Oriental Institute of the University of Chicago*, vol. 12, (Chicago: Oriental Institute, 2012), p. 157–58, s.v. *tbṱ*.

[30] For *tbṱ*, see F. Ll. Griffth and Herbert Thompson, The Demotic Magical Papyrus of London and Leiden, vol. 3 (Oxford: Clarendon Press, 1904), p.93, s.v. tbt. For *tšr*, see Joachim Quack, 'Weitere Korrekturvorschläge, vorwiegend zu demotischen literarischen Texten,' *Enchoria* 25 (1999), p. 43.

[31] Willy Clarysse and Dorothy Thompson, *Counting the People in Hellenistic Egypt*, vol. 2 (Cambridge: Cambridge University Press, 2006), p. 188.

[32] Edda Bresciani, *L'archivio Demotico del Tempio di Soknopaiu Nesos*, vol. 1 (Milan: Cisalpino/Goliardica, 1975) p. 139, n. 62 r.3.

[33] Willy Hartner, 'The Earliest History of the Constellations in the Near East and the Motif of the Lion-Bull Combat,' *Journal of Near Eastern Studies* 24 (1965): p. 12.

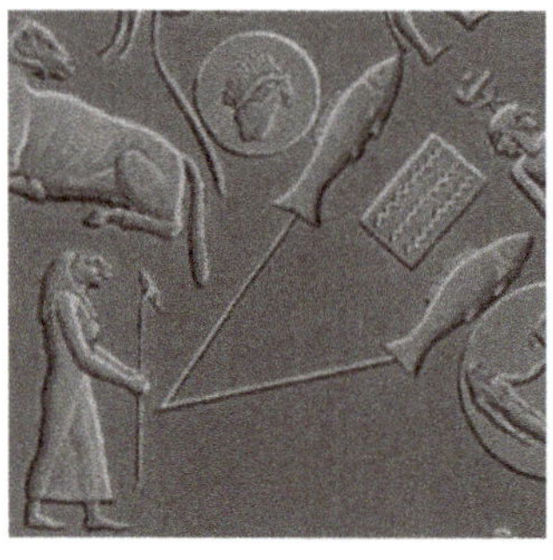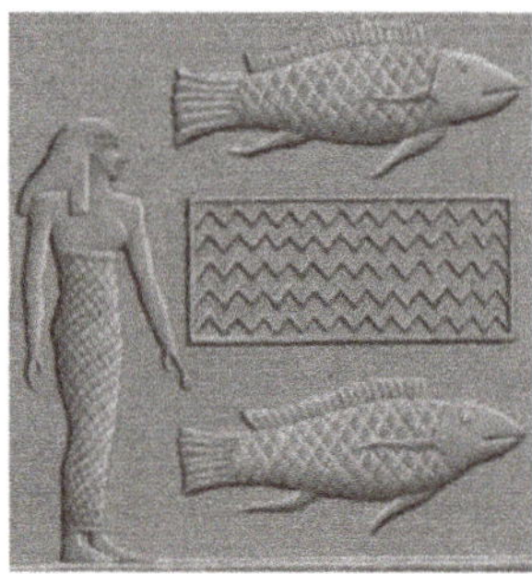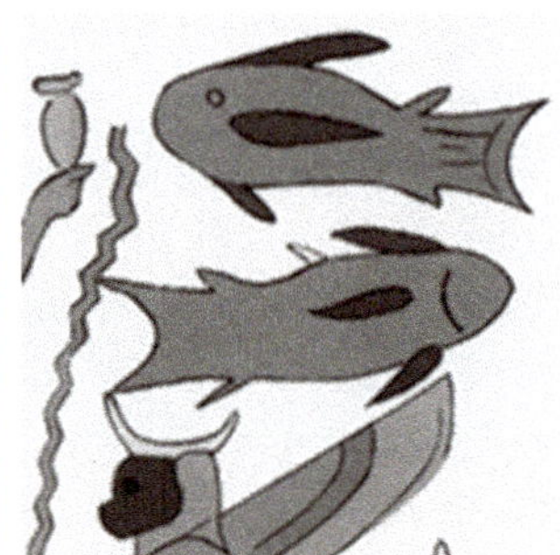

Figure 2.4 (left): Pisces from the Circular Zodiac of Dendera (Late Ptolemaic [50–30 bce], Dendera)[34]; Figure 2.5 (centre): Pisces from the Rectangular Zodiac of Dendera (Roman [20 ce], Dendera)[35]; Figure 2.6 (right): Pisces from the Tomb of Pamehit (Roman [after 148 ce], Athribis)[36]

omit the pool from the iconography. (See figure 2.6.)

The images of Pisces with a pool or well depict two fish looking in the same direction, and most Egyptian depictions without a body of water continue this arrangement. The coffin of Kleopatra and the tombs of Salamuni present two fish facing the same direction. The zodiacs of the coffin of Kornelios Pollios and the Temple of Khnum at Esna align the fish in this way, but a cord connects the heads of the fish in keeping with the practice of Egyptian fishermen, not the tails like the Mesopotamian offering-strings. (See figure 2.7.) The tombs of the Dakhla Oasis even add a fisherman holding the cord.[37] In this detail, the zodiac of the Dakhla Oasis prefigures the mosaic zodiac with Hebrew annotations at Sepphoris, but the motif of opposed fish may have produced a more fertile legacy.

As noted above, some Egyptian sources resurrect the old motif of two fish swimming in opposite directions. Both the zodiacal tombs of Athribis apply this traditionally Egyptian iconography to the Mesopotamian zodiac. The now-lost marble plaque of Daressy also counted this older motif among its depictions of the zodiacal signs. The Nile, not the Tigris or Euphrates, was home to the motif of the inverted fish, but the imagery flowed throughout the Mediterranean.

---

[34] Louvre D38; *Description de l'Egypte*, second edition, vol. 4, (Paris: Panckoucke, 1822), pl. 21.

[35] *Description*, 4, pl. 20.

[36] Flinders Petrie, *Athribis* (London: Quaritch, 1908), pl. 37.

[37] Jürgen Osing, *et al.*, *Denkmäler der Oase Dachla* (Mainz am Rhein: von Zabern, 1982), pl. 40 (and possibly 41).

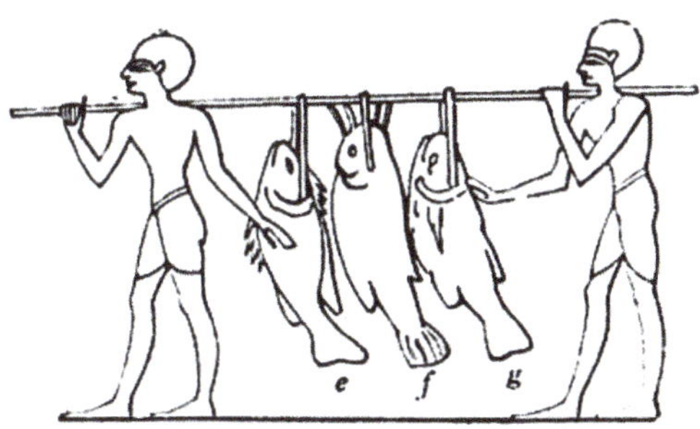

FIGURE 2.7: Typical Egyptian Use of Cords with Fish (Dynasty 5 [2500–2350 BCE], Tomb of Ty, Saqqara)[38]

Greece and Beyond

While the Egyptian inclusion of a pool or well hints at knowledge of Mesopotamian traditions, Greek sources also indicate direct awareness of Mesopotamian sources by identifying one of the 'tails' as belonging to a swallow. Aratus describes the constellation of Pisces in a mere six lines, but Theon clarifies that 'they say the more northerly of the fish, which the Chaldaeans call the Swallow, has the head of a swallow' (τοῦτον τοίνυν τόν βορειότερον ἰχθύν χελιδόνος ἔχειν τήν κεφαλήν φασιν, ὅν Χαλδαῖοι καλοῦσιν ἰχθύν χελιδόνιον).[39] This identification also appears in a horoscope for 81 CE, wherein Saturn is described as 'descending from the Swallow-Fish' (ἐπί του χελειδόνιαιου ἰχθύος καταβιβάζων).[40] Greek astronomers thus accurately preserved the species of bird which Babylonians associated with Pisces.

Classical sources also share the Babylonian ambiguity about whether the fish of Pisces were a freshwater or saltwater species. As noted above, the Babylonian word *nūnu* refers to fish generally. Classical authors differ about the environment of the fish of Pisces. Because of the proximity of Pisces to Aquarius, which he claims to be 'part of the sea' (*pars maris*), Manilius associates the fish with Neptune;[41] Ptolemy describes the fish as 'riverine' (ποτάμιον).[42] Pseudo-Proclus objects, declaring the fish to be 'oceanic' (θαλάσσια) – only to declare later their home to be 'of sweet

---

[38] J. Gardner Wilkinson, *A Popular Account of the Ancient Egyptians*, vol. 2 (London: John Murray, 1854), p. 190.

[39] Johann Gottlieb Buhle, *Arati Solensis phaenomena et diosemea*, vol. 1 (Leipzig: Officina Weidmannia, 1973), p. 63, ln. 242.

[40] Otto Neugebauer and Henry van Hoesen, *Greek Horoscopes*, Memoirs of the American Philosophical Society 48 (Philadelphia: American Philosophical Society, 1958), p. 23.

[41] Manilius, *Astronomicon*, 2.447.

[42] Ptolemy, *Tetrabiblios*, 2.7.

waters' (γλυκέων ὑδράτων).[43] These discrepancies about the environment of the fish may derive from differences in the sources which Greek authors followed, attempts to rationalize divergent traditions, or the assimilation of Mesopotamian fishes with Classical mythology.

Where Egyptians seemed content to repurpose decorative motifs vaguely linked to rebirth, Greeks strove to assimilate the fish of Pisces with mythology. Because no Babylonian myth explains the iconographic elements of Pisces, their similarity to offering-strings constitutes a modern observation. Some connection did bind the south-eastern fish to the goddess Anunitu, and Babylonian Anunitu joined syncretistically with Assyrian Astarte. Greeks, in turn, identified Astarte with Aphrodite through *interpretatio graeca*. This assimilation enabled Hyginus and Ampelius to connect the constellation to the flight of Aphrodite and her son from Typhon as fish in the Euphrates.[44] Although Ovid memorialised many divine transformations, for him the fish were merely the means of transportation.[45] Eratosthenes connected the same story to the constellation of the Southern Fish.[46] Like the Egyptian decorative element, the existence of the folkloric motif facilitated the assimilation of the zodiac in Greece.

Greek astronomers omit the Mesopotamian constellation of the Field from Pisces, but they do incorporate the iconography of the cord of the fish. Aratus draws special attention to this element by naming the star, the 'knot of heaven' (σύνδεσμον ὑπουράνιον), punningly recalling the description of the fish and 'the bonds of their tails' (δεσμά οὐραίων).[47] The wordplay of Aratus suggests a greater familiarity with the Babylonian iconography of Pisces (or, at least the name *zibbāti* [tails]) than the Egyptian adaptation. Although the association of the Cord with the Fish began in Babylon, some early Greeks treat the Cords as a separate constellation. The commentary to Aratus by Hipparchus calls the star 'the knot of the Cords' (ὁ εν τῷ συνδέσμῳ τῶν λινῶν), seemingly separating the two elements into two constellations.[48] Geminus also separates the Cords from the Fish.[49] Hyginus may explain the tendency of Hipparchus and Geminus to treat the Cords separately when he emphasizes the importance of this star as binding all the heavens, not just the fish of Pisces.[50] In fact,

---

[43] Pseudo-Proclus, *Quadripartitum*, p. 67 and 70.

[44] Hyginus, *Poeticon Astronomicon*, 2.30; Ampelius, *Liber Memorialis*, 2.12.

[45] Ovid, *Fasti*, 2.458–474.

[46] Carl Robert, ed., Eratosthenis *catasterismorum reliquiae* (Berlin: Weidmann, 1878), pp. 180–81.

[47] Aratus, *Phenomena*, 2.245.

[48] Carl Manitius, ed., *Hipparchi in Arati et Eudoxi Phaenomena Commentariorum* (Leipzig: Teubner, 1894), pp. 220 and 254.

[49] Geminus, *Isagoge*,3.7.

[50] Godefroid de Callataÿ, 'The Knot of the Heavens', *Journal of the Warburg and Courtauld Institutes*

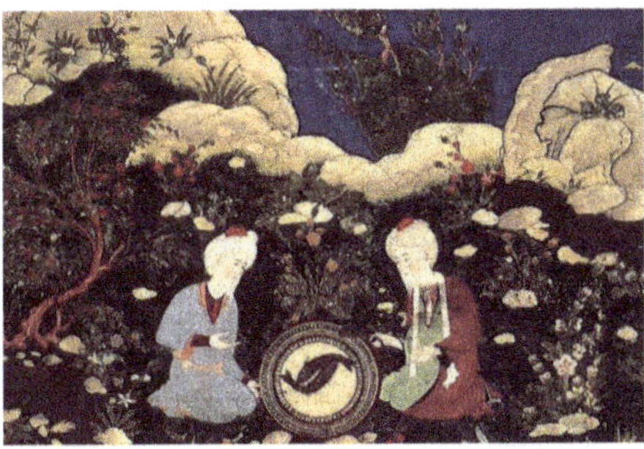

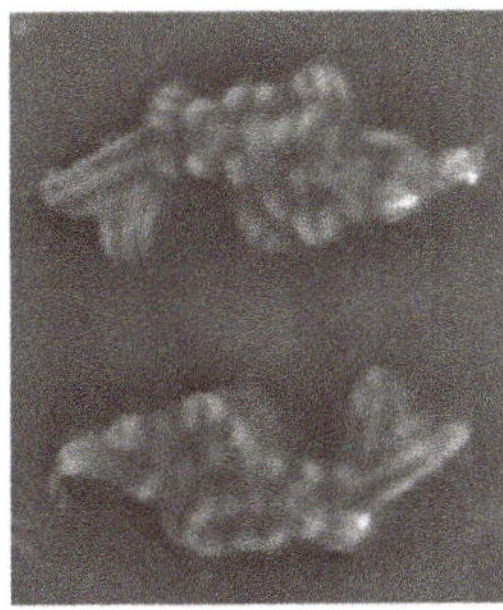

FIGURE 2.8 (left): Elijah and al-Khidr at the Well of Life (Timurid Dynasty [1370–1507], Herat, Afghanistan)[53]; FIGURE 2.9 (right): Circular Dish with Everted Lip and Twin Fish Decor, (Southern Song Dynasty [1127–1368], Lishui, Zhejiang, China)[54]

the star indicated the contemporary position of the Sun during the Vernal Equinox. In their translations of Aratus, the Latin authors Cicero, Germanicus, and Avienus retain the Cord as an iconographic element of Pisces and concur that bond holds the tails, not the heads, of the fish.[51]

The Classical tradition of uranographies follows these authors by showing two fish oriented in the same direction with tails bound by a cord, but many decorative Hellenistic depictions of the zodiac adopt the Egyptian tradition of two opposed fish. While some depictions of Pisces orient the fish in the same direction or reduce the sign to a single fish, the motif of inverted fish proliferated throughout mosaic floors, reliefs, parapegmata, ceilings, gems, coins, decorative globes, and Mithraic scenes of tauromachy. These images frequently introduced variations of this Egyptian motif in which the cord may be arranged decoratively or be absent altogether. The assimilation of the decorative motif with the zodiacal sign enabled an Egyptian pattern to permeate the Hellenistic world. Inverted fish appeared in an illustration of Elijah and al-Khidr at the fountain of life (see FIGURE 2.8), and the motif became popular in decorations of the Southern Song Dynasty of China (see FIGURE 2.9).[52] Because this eastward path roughly matches the diffusion of astrology, possibly the artistic trope was disseminated in the course of communicating astrological doctrines.[53][54]

59, (1996): p. 1–13.

[51] Cic. *Phenom. Arat.*, 249–50; Germ. *Arat.*, 244–45; Avienus; *Arat.*, 801–3.

[52] Adam T. Kessler, *Song Blue and White Porcelain on the Silk Road*, Studies in Asian Art and Archaeology 27 (Leiden: Brill, 2012), p. 141.

[53] F1937.24, Freer Gallery of Art, Smithsonian Institute, via Wikimedia Commons.

[54] JSMA 1996: 1.4, *Ferster Asian Art Collection*, Jordan Schnitzer Museum of Art, University of Oregon, Eugene, OR, longquan celadon.

Libra

Mesopotamia

The predilection of Classical sources to substitute the 'Claws of the Scorpion' for Libra has inspired frequent investigations into the origins of Libra. Classical authors report that 'ancient astronomers' preferred the term Claws to that of Libra. Separated from their Mesopotamian precursors, though, these comments prompted early modern historians to reconstruct an origin of the zodiac that differs from the Mesopotamian evidence.[55] Georg Thiele summarizes the position of classical philologists in his conclusion that

> until sometime in the first century BCE, the zodiac, which until then had only eleven images, of which Scorpio had to stretch across two signs, was completed through the introduction of a symbol for the Equinox, the Balance.

> (Etwa im ersten Jahrhundert vor Christi Geburt wurde der Tierkreis, der bis dahin nur elf Bilder hatte, von denen der Skorpion für zwei Zeichen reichen musste, durch Einführung des Symbols der Tag- und Nachtgleiche, der Waage, vervollständigt.)[56]

As long as archaising artefacts like the Brindisi disc can be read as confirmation of an eleven-sign phase of development, this reconstruction appears viable.[57] The mirage of an eleven-constellation zodiac still appears even in specialized studies.[58] Because of the pre-eminence of Classical sources, the Mesopotamian counter-evidence to this reconstruction of the zodiac still struggles for recognition.

The connection between the Claws and Libra is not as remote as Classical sources suggest. Early evidence separates the stars of Scorpio from the constellation which became Libra. 'Astrolabe B' assigns the Balance (^{mul}zi-ba-ni-tum, written in Akkadian, contrary to the predominant Sumerograms) to the middle section of the horizon during Month VII. This early reference to the Balance stands next to Scorpio (^{mul}GÍR.TAB[$=zuqaqīpu$]) which occupied the same section of the horizon in Month

---

[55] George Costard, *The History of Astronomy* (London: Lister, 1768), p. 19.

[56] Georg Thiele, *Antike Himmelsbilder* (Berlin: Wiedmannsche Buchhandlung, 1898), p. 70.

[57] Karl Kerényi, 'Die religionsgeschichtliche Einordnung des Diskos von Brindisi', *Mitteilungen des Deutschen Archäologischen Instituts. Römische Abteilung* 70 (1963): pp. 93–99, pl. 42–43; Karl Kerényi, *Dionysius* (Princeton: Princeton University Press, 1976), pp. 385–86; cf. Hans Georg Gundel, *Zodiakos: Tierkreisbilder im Altertum*, Kulturgeschichte der antiken Welt 54, (Mainz am Rhein: Philipp von Zabern, 1992), p.71–72, 237.

[58] Bruce Underwood, *Ordering the Heavens* (Leiden: Brill, 2007), pp. 221–22.

VIII. However, as early as MUL.APIN (before 687 BCE), a scribe had connected the two constellations by writing '^{mul}ZI.BA.AN.NA[=*zibānītu*] SI[=*qarnu*] ^{mul}GÍR.TAB' (the constellation *Zibānītu* is the pincers of the scorpion).[59]

The scribe who explained the relationship between Libra and the Claws intended to resolve a syntactical ambiguity. When Mesopotamian astronomers chose the constellations by which they referred to astral phenomena and standardized their names, they may have derived the names of two zodiacal signs, Pisces and Libra, from the same root, *zbn* (tail or rear part). In the case of Pisces, *zibbāti* constitutes a feminine plural and may be translated as 'the Tails'. In the case of Libra, *zibānītu* comprises a feminine dual and is restricted in sense to a type of balance ('provided with two **zibana*').[60] The two terms signify distinctly different objects, but their signifiers are distinguishable only by grammatical number. In order to distinguish between the two terms, the scribe introduced a description which referred to another zodiacal sign.

Whereas *zibbāti* referred to the 'tails' of either a swallow or a fish, the type of balance invoked by *zibānītu* remains somewhat obscure. Lexical lists compiled by Mesopotamian scribes endeavouring to clarify old Sumerian terms and collect the lexemes of their own language illustrate the cuneiform terminology of scales. For a simple scale, these scribes wrote *gišrinnu*, a loanword from the Sumerian GIŠ. ERÍN (composed of the graphemes for 'wooden tool' and 'warehouse'). In these lists, scribes distinguished the *zibānītu* as a subset of scale, 'a scale provided with two **zibana*', with **zibana* possibly being a Sumerian loanword. Because scales contain relatively few parts, the semantic range expressed by citations of *zibānītu* may clarify its meaning. In one poem, both dishonest and honest merchants hold the *zibānītu*.[61] Some scribes write *zibānītu* as ^{giš}*zibānītu*, a writing which prefixes the determinative for wooden tools. Lexical lists describe the *zibānītu* as having two parts: the 'arm of the *zibānītu*' (*aḫu ša zibānīti*) and the 'leather bag of the *zibānītu*' (*luppu ša zibānīti*). Another scribe wrote about the 'shallow bowls' (*itquru*) of the *zibānītu*. The balance of these citations suggests that the *zibānītu* consists of the crossbeam (*aḫu*) and the parts hanging from it like two tails (*zibānītu*).

In the course of compiling the observations which constituted the *Astronomical Diaries* and precipitated the transition from zodiacal constellations to zodiacal signs, Mesopotamian astronomers changed the name of Libra. Contrary to the reports of Classical authors, these Mesopotamian observers replaced *zibānītu* with RÍN

---

[59] Hunger and Pingree, *MUL.APIN*, p.33, ii.11.

[60] Gelb, *Assyrian Dictionary*, s.v. *zibānītu*.

[61] Wilfred Lambert, *Babylonian Wisdom Literature* (Oxford: Clarendon Press, 1960) p.132, ln. 107, 111.

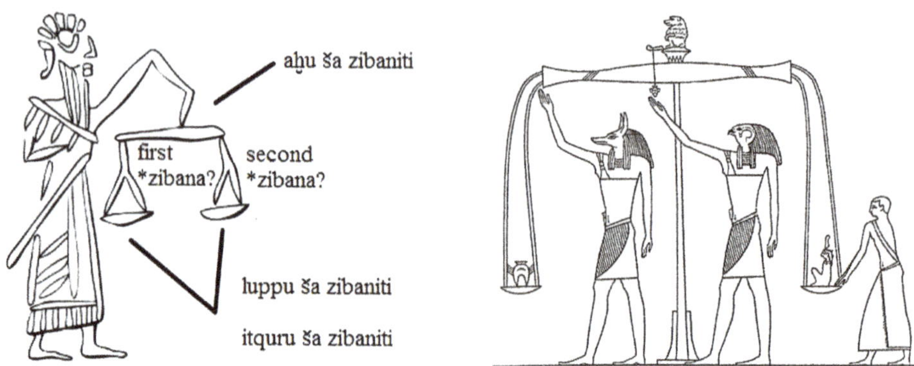

FIGURE 2.10 (left): *Zibānītu* from a Cylinder Seal (2350–2100)[62]; FIGURE 2.11 (right): Weighing of the Heart Vignette (Ptolemaic [330–330 BCE], Egypt)[63]

(=*gišrinnu*), a more general term for a balance. Like the combination of the Swallow and the Fish into ^mulKUN.ME, this change appears in the earliest *Astronomical Diary* from 652 BCE, but other contemporary compositions also discuss specific stars of ^mulGIŠ.RÍN.[64] Notwithstanding this modification, *zibānītu* survives into the Hellenistic era.[65] However, 'the Claws' also survives, appearing in a horoscope for 410 BCE.[66] Thus, by the time Alexander encountered the Mesopotamian zodiac, Libra lay claim to three names: the long-standing *zibānītu*, its clarification SI ^mulGÍR.TAB, and the astronomically preferred RÍN.

Egypt

Like fish, the balance formed an important element of Egyptian iconography long before the standardization of the zodiac. Whereas depictions of weighing rarely appear in Mesopotamia, they were common in Egypt. One of the most famous

---

[62] Detail from an unknown cylinder seal or sealed tablet. Cf. Jeremy Black and Anthony Green, *Gods, Demons, and Symbols of Ancient Mesopotamia: An Illustrated Dictionary* (Austin: University of Texas Press, 1992), p. 183.

[63] Museo Egizio 1791 (TM 57201), Achille Prisse D'Avennes, *Histoire de l'art égyptien d'après les monuments depuis les temps les plus reculé jusqu'à la domination Romaine*, Atlas, vol. 2 (Paris: Bertrand, 1878), pl. 8.

[64] Sachs and Hunger, *Diaries*, p. 42; David Pingree and Christopher Walker, 'A Babylonian Star Catalogue: BM 78161', in Erle Leichty, Maria deJong Ellis, and Pamela Gerardi, eds., *A Scientific Humanist: Studies in Memory of Abraham Sachs*, (Philadelphia: University Museum, 1988), p. 315.

[65] Arthur Ungnand, 'Besprechungskunst und Astrologie in Babylonien', *Archiv für Orientforschung* 14 (1941), p. 257.

[66] Rochberg, *Horoscopes*, p.57.

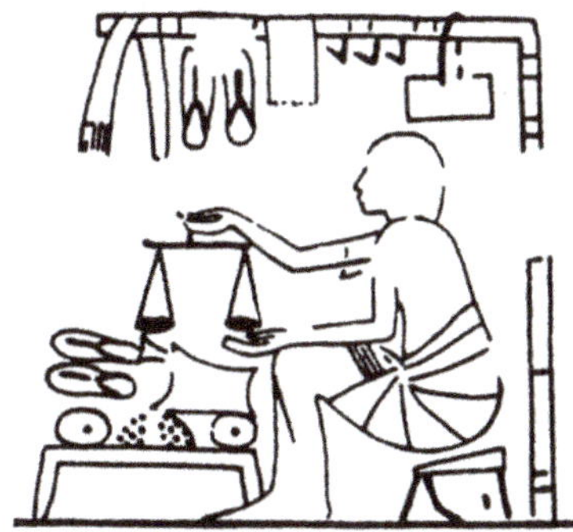 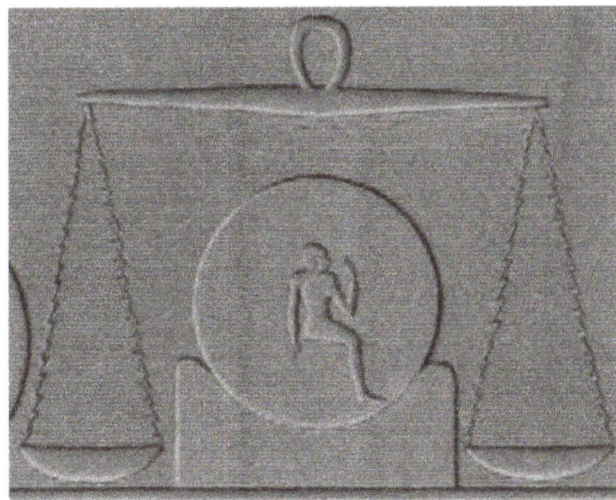

Figure 2.12 (left): An Egyptian Hand-Balance (Amenhotep III [1391–1351 BCE], Thebes)[67]; Figure 2.13 (right): Libra from the Rectangular Zodiac of Dendera (Roman [20 CE], Dendera)[68]

Egyptian images is the 'Weighing of the Heart' vignette from the *Book of the Dead*. In these vignettes, the crossbeam of the scales is mounted on a stand and occasionally equipped with a level. (See Figure 2.11.) In a handful of depictions of mundane measurements, the crossbeam is held by the measurer. (See Figure 2.12.)

The balance even served as a hieroglyph (U38, *mḫꜣ.t*, balance), as did its parts.[69] The post of the balance (U39 and U40) could denote the post of the balance (*wṯs.t*) or function as the determinative of verbs of lifting, like *wṯs* or *ṯsy*.[70] The crossbeam (U91) saw limited use as a hieroglyph but appears as an ideogrammatic determinative with no phonetic value in *iwsw* (hand balance), *mḫꜣ.t* (balance), *ḥmꜣg* (balance), and *ḥnkw* (the pans of the balance).[71] The Egyptian name of Libra, though, does not derive from any of these terms.

Like Pisces, the name of the zodiacal sign of Libra requires interpretation. In Egypt, Libra first appears in a Demotic text, Strasbourg D 521, which lists zodiacal signs and the corresponding Egyptian month.[72] The name for Libra is not the common Demotic word for scales, *mḥy.t*, which has a clear morphology as a feminine, *m*-substantive of the verb *ḫꜣy* (measure), with a hieroglyphic precursor *mḫꜣ.t* and

---

[67] TT 162, cf. N. de Garis Davies and R. O. Faulkner, 'A Syrian Trading Venture to Egypt', *Journal of Egyptian Archaeology* 33 (1947): pp.40–46, pl. 8.

[68] Denon, *Description de l'Egypte*, vol. 4, pl. 40.

[69] Gardiner, *Grammar*, p. 521.

[70] Gardiner, *Grammar*, p. 520.

[71] Rainer Hannig, *Grosses Handwörterbuch Ägyptisch-Deutsch*, Kulturgeschichte der Antiken Welt 64 (Mainz: Philipp von Zabern, 2006), p. 1448.

[72] Otto Neugebauer, 'Egyptian Planetary Texts', *Transactions of the American Philosophical Society*, n.s. 32, part 2 (1942): p. 246.

Coptic descendants in various dialects as μαϣε, μαϣι, or μαϩε. Instead, Strasbourg D 521 introduces a different word: *iḥy.t*.

Wilhelm Max Müller and Wilhelm Spiegelberg had debated the reading and translation of Strasbourg D 521 before Otto Neugebauer dismissed their debate in a footnote and joined his reading of the sign to the date which he proposed for D 521:[73]

> Accepting this conclusion [= a date around 250 BCE], a problem might find its solution which otherwise seems to be without possibility of explanation, namely the denomination of the sign ⌴ as 'horizon' (*ȝḥ.t*) in all our Demotic sources,[23] known to us as the 'balance'.[74]

Although the date could be challenged on palaeographical grounds, the question of whether astronomy verifies palaeography or palaeography confirms astronomy is an epistemological one. Fortunately, an absolute date is not germane to an etymology. What is more important is the fact that when Neugebauer wrote, the etymology of the Egyptian term for Libra was in doubt. In his footnote, Neugebauer dismisses two somewhat desperate attempts by Müller to restore a prothetic *m* to *iḥy*.[75] Neugebauer overstates the evidence by claiming that Libra appears as horizon (*ȝḥ.t*) in every Demotic source. Because the passage addresses Strasbourg D 521 directly, it may be presumed to be an outlier. Not only does Neugebauer transliterate this word as *iḥy.t* in Strasbourg D 521, but when he tabulates the palaeographical variants of Libra, two ostraca from Medinet Habu also preserve *iḥy*, as does a particularly unambiguous writing in P. Cairo 50143 (TM 48726).[76] These last three instances perversely introduce the word with a feminine article while omitting the suffix *–t*, characteristic of Egyptian feminine nouns.

After discussing Greek and Mesopotamian terminology, Neugebauer returns to the

---

[73] Wilhelm Spiegelberg, 'Ein aegypisches Verzeichnis de Planeten und Tierkreisbilder', *Orientalistische Litteratur-Zeitung* 5, no.1 (Jan. 1902): col. 6–9; W. Max Müller, 'Zu dem neuen Strassburger astronomischen Schultext', *Orientalistische Litteratur-Zeitung* 5, no.4 (Apr. 1902): col. 135–36; W. Max Müller, 'Zur Geschichte der Tierkreisbilder', *Orientalistische Litteratur-Zeitung* 6, no. 1 (Jan. 1903): col. 8–9; Wilhelm Spiegelberg, 'Die ägyptischen Namen und Zeichen der Tierkreisbilder in demotischer Schrift', *Zeitschrift für Ägyptische Sprache und Altertumskunde* 48 (1910): p.146–51.

[74] Neugebauer, 'Horoscopes', p.122.

[75] Müller, 'Schultext', col. 135; Müller, 'Geschichte', col. 8; Neugebauer, 'Horoscopes', p. 121; Herman Grapow, *Über die Wortbildung mit einem Präfix m- im Ägyptischen*, Abhandlungen der Preussischen Akademie der Wissenschaften, Philosophisch-historische Klasse 1914 no. 5 (Berlin: Verlag der Königliche Akademie der Wissenschaften, 1914), p. 27.

[76] Neugebauer, 'Horoscopes', p.121 and pl. 3.

problem of the Egyptian name of the zodiacal sign and its disparity with Egyptian depictions:

> The latter statement [= that Hellenistic cuneiform separates RÍN from GÍR.TAB] holds equally for the Egyptian representations of the zodiac, the earliest preserved examples being at Dendera (time of Tiberius). In the pictorial representations a clear balance is always given and nothing like a "horizon." How can this contradiction be explained?[77]

Here, Neugebauer notes that the object signified by the Egyptian term for Libra differs from the object depicted in the iconography, but he again overstates the evidence by claiming that nothing like a horizon was depicted. In fact, the rectangular zodiac of Dendera depicted the crossbeam of the scales over a hieroglyphic representation of the horizon. (See FIGURE 2.11.) Most written instances of Libra reduce the sign to a hieroglyphic depiction of the sun on the horizon (N 27). Despite his dismissal of these writings, Neugebauer derived his etymology from this hieroglyph, vocalized as *3ḫ*.

Neugebauer did not attempt to derive *3ḫ.t* from an *i*-prefix or some other morphological manipulation of the root *ḫ3y* (measure).[78] Rather, he attributed the etymology of *iḫy* to the root *3ḫ* (glory), a reading derived from the connotations of the biliteral hieroglyph N 27. In previous phases of the Egyptian language, the root *3ḫ* had generated *3ḫ.t* (horizon), and Neugebauer bases his etymology on this form of the root. Despite the fact that horizons commonly appear in Egyptian metaphors even into Demotic, no clear instance of the word for horizon appears in Coptic. The root *3ḫ* survives as εοου (glory), with an adjectival form Ϩαε (glorious). Neugebauer does not address the morphology of *iḫy*, but several Demotic references to the horizon do substitute *i* for *3*.[79] Neugebauer begs the question, though, when he tries 'to discover a reasonable motive for calling a zodiacal sign "horizon"[80]' and justified the reading with astronomical phenomena and astrological doctrines:

> From this assumption [= the equation of the first month with Scorpio] follows, namely, that the preceding sign 'balance' was rising heliacally at the beginning of the year – sufficient reason, indeed, to be called '(being in the) horizon'. Such an emphasis on the quality of a constellation to indicate the beginning of the

---

[77] Neugebauer, 'Horoscopes', p. 122.
[78] Gerhardt Fecht, *Wortakzent und Silbenstruktur: Untersuchungen zur Geschichte der ägyptischen Sprache*, Ägyptologische Forschungen 21 (Glückstadt: J.J. Augustin, 1960), p. 6.
[79] Johnson, *Dictionary*, vol. 11, s.v. *iḫy(.t)*.
[80] Neugebauer, 'Horoscopes', p. 122.

civil year by its heliacal rising would certainly be nothing surprising in Egypt; moreover this sign is, so to speak, the 'Horoscope' of the year.[81]

Not only is this justification astronomically untenable, but the reasoning depends on an anachronistic understanding of the development of astrology. Footnote 29 in the citation notes that decans are called the 36 'horoscopes' in P. Lond. 98, but the constellation of Libra did not occupy the 'Horoscope' or ascendant on the first day of the year. Neugebauer is correct that these stars made a heliacal rising on this day, but this phenomenon fixes their position in the twelfth astrological place, not the first. The invocation of '(being in the) horizon' might be saved by reference to the Egyptian phrase pr m 3ḫ.t (coming forth from the horizon) but another more contemporary phrase for a heliacal rising, ḫʿ (appearance) dominates Demotic astronomical texts and does not refer to the horizon.

The connection between the horizon and the Egyptian name of Libra is probably not worth pursuing. First, if the etymology of this reading of Libra is coupled with the date of 250 BCE proposed by Neugebauer, Strasbourg D 521 marks the earliest acknowledgment of the ascendant. Secondly, Neugebauer does not explain why this association persisted after 130 BCE, when it was no longer astronomically correct. Granted, the term for Libra reverted to μαϣε in Coptic, but instances of *iḫy*, albeit written ideogrammatically and thus perhaps pronounced 3ḫ.t, appear in Demotic until at least 195 CE.[82] By this time, though, the Egyptian calendar began with the Sun in Cancer. While the ascendant probably did originate in Egypt, the use of the ascendant proposed by Neugebauer would constitute the earliest example of katarchic horoscopy and the earliest use of 'Great Year' astrology, both of which appear in Greek sources but are more strongly associated with regions east of Egypt.[83] Some connection might be postulated between the first season (3ḫ.t) of the year and the sign of Libra, but Neugebauer does not explore this connection and it suffers from the same shift in seasons. A more promising connection lies in the fact that every other name of a zodiacal sign referred to its iconography.

An attempt to explain *iḫy* as a Demotic borrowing from Akkadian falls short of satisfying Winand's criteria for a Semitic loanword. Although *iḫy* is often written phonetically, many instances preserve writings as ideograms. Secondly, while *iḫy*

---

[81] Neugebauer, *Horoscopes*, p. 122.

[82] Micah Ross, 'A Provisional Conclusion to the Horoscopic Ostraca of Medinet Madi', *Egitto e Vicino Oriente* 34 (2011): pp. 49–51.

[83] Dorian Gieseler Greenbaum and Micah Ross, 'The Role of Egypt in the Development of the Horoscope', in *Egypt in Transition: Social and Religious Development of Egypt in the First Millennium* BCE (Prague: Czech Institute of Egyptology, 2010), pp. 146–82.

did not appear in earlier Egyptian texts, the ideogram did and thus the word has an etymological derivation within Egyptian, albeit an impenetrable one. On the other hand, *iḥy* does have a Semitic cognate that can be validated at (morpho-)phonological and semantic levels. The word *aḫu*, identified as the crossbeam of the Mesopotamian balance, refers to the object shown in the Egyptian images. The phonetic writing approximates the Akkadian central open vowel /a/ with a grapheme for the Demotic closed front vowel /i/. However, from Middle Egyptian to Coptic, many words shifted from /i/ to /a/.[84] Consider, for example, the phonetically similar interrogative *iḥ*. Coptic renders this word as *aⲩ* in the Sahidic and Bohairic dialects, and *ⲉⲩ* in Fayyumic, and *aⳉ* in Amharic. In both Demotic and Akkadian, the second consonant is the same and the final vowels are weak in both phonological systems.

Notwithstanding the phonological modifications between Akkadian and Demotic, a stronger objection to the interpretation of *iḥy* as a loanword is the fact that despite a deliberate change in the name of Libra by Mesopotamian astronomers, the word *aḫu* never actually appears as the name of Libra. Against this objection, Egyptians more often depicted Libra as a simple crossbeam rather than the established motif of the crossbeam and scales mounted on a post. On the balance, though, the identification of *iḥy* as a Demotic rendering of *aḫu* is tentative and buoyed more by the clear Mesopotamian origin of the zodiac and a dissatisfaction with the existing etymology. Although the proposals by Neugebauer may be rejected, the derivation of *iḥy* from Mesopotamian sources is far from conclusive. Ultimately, the determination of loanwords is not an exact science, and the question remains open. Nonetheless, the hypothesis of an Akkadian loanword conforms nicely to the Greek descriptions of Libra.

Greece

The foregoing discussion of the Egyptian word for Libra would be unwarranted had the Greek adaptation of the Babylonian constellation, whether it should be called *zibānītu* or *gišrinnu*, proceeded without complication. In the case of Libra, the Classical predilection to replace the zodiacal sign with the Claws has distracted most analytical discussion of the iconography of the sign. For example, Auguste Bouché-Leclercq, who remains an authority on zodiacal iconography despite the elucidation of Mesopotamian origins and a century of scholarship, devotes a mere

---

[84] William Albright, 'The Principals of Egyptian Phonological Development', *Receuils de Travaux* 40 (1923), p. 67.

thirteen lines and a footnote to the iconography of the Balance.[85] Even this brief survey focuses on the possibility of Libra as a metaphor for the equinoxes and the association of the sign with the Classical iconography of Justice.

Greek authors, however, seemingly did not simply choose the most obvious term for scales as the name of the zodiacal sign. In Greek literary texts, by far the most common word for the instrument used in weighing is τάλαντον. Other terms also exist. The related terms πλάστιγξ and its diminuitive πλαστίγγιον synecdochically refer to the balance by the scale which compared the weights; τρυτάνη appears in a minority of writers. The lexicographer Hesychius reports that the crossbeam of a balance was called the σηκωτήρ.[86] Two other words appear in single authors as well. Eustathius refers to the ζυγοστάτημα; Photius embellished the uncommon τρυτάνη into the *hapax legomenon* ζυγοτρυτάνη.[87] Despite the distinct preference of Greek literati to use τάλαντον in reference to the balance, Hellenistic astronomers unanimously describe the zodiacal sign as ζυγός (yoke). To be sure, ζυγός frequently occurs in conjunction with τάλαντον, but the reference is limited to the crossbeam of the scales – precisely the term suggested by the Akkadian *aḫu*, yet otherwise unidentified in Egyptian vocabulary, and probably the object denoted by *zibānītu*.

Images of this crossbeam (and its pans) constitute the first of three Greek strategies for the depiction of Libra.[88] However, as the sole inanimate object among the zodiac, Libra could scarcely be described by the Aristotelian term ζῳδιακός (little animal). The second strategy assimilated Libra with the Claws of the Scorpion, even to the point of reducing the number of zodiacal signs to eleven, and reveals an awareness of the Mesopotamian traditions about Libra.[89] Early Greek astronomers preferred the Claws to the Balance. Aratus and Eratosthenes exclusively refer to the Claws. Hipparchus eschewed the Claws one time in preference for ζυγός, a decision which prompted Manitius to flag the line as suspect.[90] Ptolemy also referred to ζυγός only once – when he was quoting Hipparchus for an otherwise lost observation.[91] Lest the constancy of Hipparchus suffer, it should be noted that

---

[85] Auguste Bouché-Leclercq, *L'astrologie grecque* (Paris: Leroux, 1899), pp. 141–42.

[86] Hsch., s.v. σηκωτήρ.

[87] Eust., *Il.* 665.29; Phot. *Bibl.* s.v. ζυγός.

[88] Gundel, *Zodiakos*, pp. 71–72.

[89] Gundel, *Zodiakos*, p. 42–43, 92, 100, 103,162–63, 232, 237, 249, 310, 312, 328. Specifically, the sources which Gundel counted as Nr. 29; 78,1; 95; 158, 388; 393; 457 and 460 lack Libra.

[90] Hipparchus, 3.1.5.2; Karl Manitius (ed.), *In Arati et Eudosci Phaenomena commentariorum libri tres* (Leipzig: Teubner, 1894), p. 327, s.v. ζυγός.

[91] Ptol. *Alm. 9.7.*

observation in question occurred on 262 BCE, well before the birth of Hipparchus or Ptolemy. So much for the preferences of the ancient astronomers.

The third strategy preserved a connection to living creatures by placing the crossbeam in the hand of a human figure. Whereas Egyptian artists placed the crossbeam in the hands of a colourless god with neither name nor distinguishing iconography, Greek authors and artists often attempted to identify the figure bearing the scales. For many Greek astrologers, Libra was 'masculine' because it was the seventh sign and Pythagoras had declared odd numbers to be masculine. Consequently, Firmicus describes the figure holding the scales as a man.[92] Manilius placed the Balance in the hands of its legendary male inventor, Palamedes.[93] Ampelius identified the scale-bearer with Mochos, another mythical male inventor of the instrument.[94] In other cases, Greeks placed Libra in the hands of Virgo. This association enabled the identification of Virgo with Astraea, who lays claim to the scales as the iconography of justice.[95] Even though Hyginus acknowledges the assimilation of Libra and Scorpio of the second strategy, his progression directly from Virgo to Scorpio rests equally on the assimilation of the iconography of Libra to Virgo through the iconography of justice.[96] This identification with extant motifs even directed astrological doctrine. Although Manilius kept Libra and Virgo distinct by assigning Libra its own figure, he combined the figure of justice and the yoke of servitude in his astrological dogma.

Conclusion

Mesopotamians used similar words to refer to Pisces and Libra. In the case of Pisces, this word described the tails of a bird and a fish, which had been combined from previously independent constellations in the course of the development of zodiacal signs. In the case of Libra, this word described the mechanism used for weighing. A hypothesis may be advanced that when these signs travelled from Mesopotamia to Egypt, Egyptian scribes adopted different strategies of assimilation. In the case of Pisces, the scribes may have adopted the Mesopotamian name as a loanword for a previously unknown type of fish, to which they ascribed long-standing decorative iconography. In the case of Libra, the scribes may have referred to the name of the Mesopotamian constellation by synecdoche, clarifying this reading by a modification of the Egyptian motif of weighing. While these patterns of borrowing from

---

[92] Firm. *Math.* VIII 4.7.

[93] Manilius, *Astronomicon* 4.203–16.

[94] Ampelius, *Liber Memorialis*, 2.7.

[95] Arat. 98, 105; Luc. 9.535.

[96] Hyginus, *Astronomica* 2.25–26.

Mesopotamia to Egypt remain speculative proposals, they accord with the Greek evidence. In the case of Pisces, the Greek descriptions reveal an understanding of the Mesopotamian precursors and the iconographic legacy roughly corresponds to the spread of Hellenic astrology. In the case of Libra, the choice of Greek words agrees with the Egyptian emphasis on the arm of the Balance. The linguistic evidence for the linguistic borrowing of Pisces is stronger than that for the borrowing of Libra, but the conventional etymology for the Egyptian rendering of Libra is astronomically untenable.

Acknowledgements
Thanks are due to Prof. Michio Yano who questioned why some zodiacs showed the fish of Pisces pointing in the same direction but others portrayed these fish facing opposite directions and to the Sophia Centre for its sustained support of analyses of the development of the iconography of the zodiac in Mesopotamia, Egypt, Greece, and India. Further thanks are due to Dorian Gieseler Greenbaum, Stephan Heilen, Franziska Naether, and Naoko Wolze.

Dwarfs as an Ancient Mayan Metaphor for the Stars

Suzanne Nolan

ABSTRACT: Ancient Maya artists commonly embedded hieroglyphics into pictorial art to help define and inform readings of the iconography. On Hieroglyphic Stairway 2 (HS. 2), at Yaxchilan, this can be seen in the form of two dwarfs bearing the *ek'* hieroglyph on their bodies. Montgomery (2006) defines it thus: EK' (ek'/Ek') (T510af) 1> n. 'star'; represents one-half of the full 'star' glyph 2> n. 'Venus'? The word for 'dwarf', *aak*, has also been identified as the 'turtle constellation', because of the homophony between that and the word for 'turtle', *ahk*: either Gemini (Roys 1965; Milbrath 2000) or Orion (Foster 2005). Why have the dwarfs of HS2 been qualified with the *ek'* hieroglyph? What can this tell us about their identities and purpose on HS. 2? This chapter seeks to answer these questions, and discuss how the dwarf could be used as a metaphor for a celestial body – whether a constellation or planet – within Maya art and architecture. It will demonstrate the multi-layered meaning of the dwarfs' presence on HS. 2, as metaphors for astrology and astronomy in Maya art.

Period	Division	Dates
Preclassic	Early Preclassic	2000–1000 BCE
	Middle Preclassic	1000–350 BCE
	Late Preclassic	350 BCE–CE 250
Classic	Early Classic	CE 250–550
	Late Classic	CE 550–830
	Terminal Classic	CE 830–950
Postclassic	Early Postclassic	CE 950–1200
	Late Postclassic	CE 1200–1539

TABLE 3.1: Maya Chronology[1]

Introduction

This chapter explores the role of the dwarf in Classic period Maya culture, and discusses the way in which it was used as a representation for celestial bodies, among

---

[1] Adapted from Francisco Estrada-Belli, *The First Maya Civilization: Ritual and Power before the Classic Period* (Abingdon, UK: Routledge, 2011), 3.

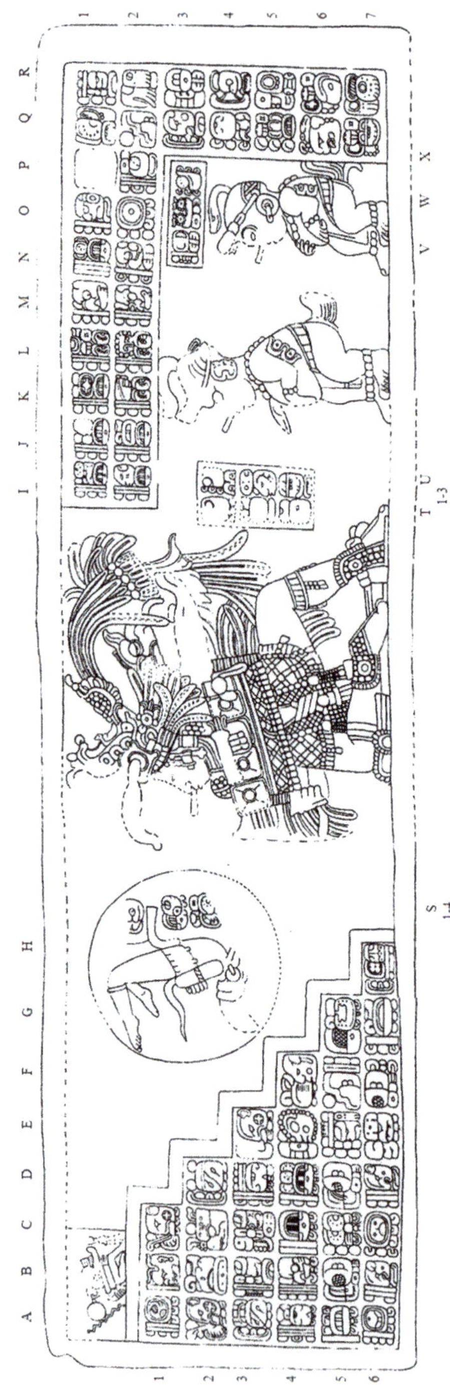

Figure 3.1: HS. 2 Step VII, drawing by John Montgomery, FAMSI drawing collection #JM01604. Copyright © 2000 John Montgomery.

other things, in Maya art. It will also discuss the significance of dwarfs in ancient Maya imagery with particular reference to one monument at the archaeological site of Yaxchilan: Step VII of Hieroglyphic Stairway 2 (henceforth referred to as HS. 2; FIGURE 3.1).

Understanding Maya art, or the images of any culture, requires a cultural empathy that we, as scholars, do not necessarily possess.[2] In the case of HS. 2, dwarfs are represented not only through their physical appearance on the monument, but through the signs and symbols used to allude to their physical forms, and metaphorical and intangible natures; their presence within the context is not just about *what* they represent, but the wider connotations and ideologies associated with them that we, as removed in space and time from that context, will never be able to fully comprehend. For example, when we see an image of a dwarf, our cultural experiences provide context: perhaps fairy stories of Snow White, where dwarfs are intrinsically linked with benevolence, for taking pity on Snow White and allowing her to stay with them, and caring for her after she fell asleep. Equivalent myths and histories about dwarfs in Classic period Maya thought would have called to mind particular connotations and concepts for the audience when they considered the monument (perhaps fertility, divinity, and support, to be discussed below) that we are unable to fully appreciate. Many monuments, bark-paper books (codices), and other hieroglyphic texts have been destroyed, particularly since the Terminal Classic and Postclassic periods, and the Spanish conquest. These sources would have contained within them further clues to understanding Maya ideology and culture that we, as scholars, are not able to draw from in our understanding. As a result, we must piece together ancient Maya histories and belief from the small selection of artefacts that remain to us.

We know that Maya art incorporates a huge number of visual metaphors to layer meaning and ideology. As Andrea Stone and Marc Zender make explicit, 'the extensive use of metaphor requires that the viewer be educated in the Maya system of metaphorical reference in order to grasp the imagery on all of its levels.'[3] Therefore, while we can make a 'best guess' at meaning, scholars of the ancient Maya are forced to concede an incomplete understanding of the images and texts (also laden with metaphor) that have been uncovered. The use of metaphor in Maya art and writing

---

[2] Simon Martin, 'On Pre-Columbian Narrative: Representation Across the Word-Image Divide', in J. Quilter and M. E. Miller, eds., *A Pre-Columbian World* (Washington, DC.: Dumbarton Oaks, 2006), pp. 55–61.

[3] Andrea Stone and Marc Zender, *Reading Maya Art: A Hieroglyphic Guide to Ancient Maya Painting and Sculpture* (London: Thames and Hudson, 2011), 23.

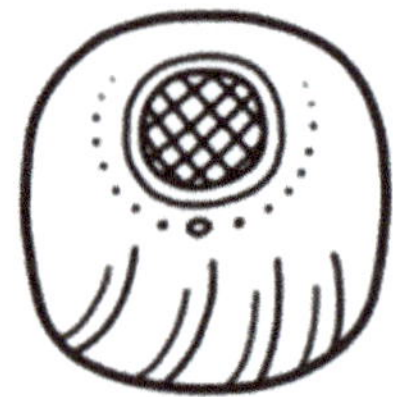

Figure 3.2: HA' *ha'* meaning 'water', 'lake', or 'river'

can be illustrated by the use of the water lily to mean 'water'. The glyph for 'water' is *ha'* which is drawn as a water lily blossom (FIGURE 3.2). In pictorial art, the water lily blossom is used in contexts to suggest fresh water, such as on the Maya 'Resurrection Plate'. Water lily blossoms flank the turtle (which represents earth) to indicate the primordial seas on which the earth floats.[4] It is possible that water lilies were used as a metaphor for clean, pure water, as well as abundance,[5] along with other cultural associations.

'Reading' Maya art also requires an understanding of the Classic Maya written language, which is a derivation of Ch'olti.[6] It includes both a syllabic (signs for syllables) and logographic (signs for whole words or concepts) signs. Logographs were often incorporated into pictorial art to add, change, or layer meanings of images. An excellent illustration of this is Stela 1, from Bonampak (FIGURE 3.3), where the Maize God emerges from a personified mountain.[7] The mountain takes the form of a zoomorphic head, but there are logographic markers identifying it as *witz*, or 'mountain': the 'hanging grape' motif, the dotted curls, and the beaded circles, all of which are found within the hieroglyph for *witz*. These indicators are used to adorn the 'face' from which the Maize God emerges to symbolise that it is, in fact, a mountain.

Gods can be similarly recognised. K'awil, for example, is a multifaceted god with associations with lightning, fertility, and serpents (among others). His name glyph is a head with an obsidian mirror in his forehead from which smoke emanates. In Maya imagery, the smoke alone can be used to represent him, and rulers impersonating him would simply wear a torch upon their forehead.[8] In order to 'read' Maya

---

[4] Stone and Zender, Reading Maya Art, 23.

[5] Lisa J. Lucero, 'Water Control and Maya Politics in the Southern Maya Lowlands', in E. A Bacus and L. J. Lucero, eds., *Complex Polities in the Ancient Tropical World* (Arlington, VA.: American Anthropological Association, 1999), 41.

[6] Stephen Houston, John Robertson, and David Stuart, 'The Language of the Classic Maya Inscriptions', *Current Anthropology* 41, no. 3 (2000): pp. 321–56.

[7] Stone and Zender, *Reading Maya Art*, 138–39.

[8] Stone and Zender, *Reading Maya Art*, p. 49.

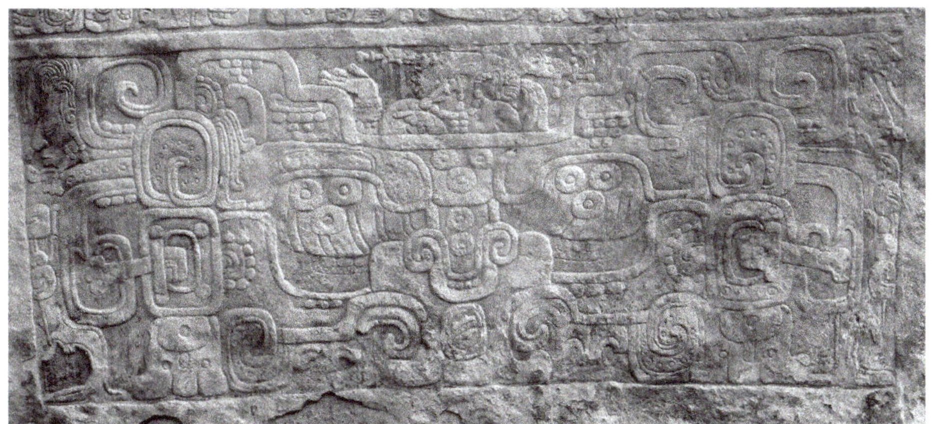

FIGURE 3.3: Basal register of Stela 1, Bonampak, showing the personified mountain incorporating details of the glyph *witz*. Author's photograph.

art, then, scholars need an understanding of hieroglyphic writing, in order to recognise where embedding, qualifying, conflation, and personification is used to deepen the meaning of the image.

This brief explanation of understanding Maya art and metaphor will be used to shape the analysis of the dwarfs that appear on HS. 2, at Yaxchilan. The geographical and temporal context in which HS. 2 was produced is also important in understanding its meaning. A short summary of this context, along with discussion of the ruler who commissioned the monument, Bird Jaguar IV, will follow.

Yaxchilan and Bird Jaguar IV

Yaxchilan rose to prominence in the Late Classic era during the reign of Shield Jaguar II, who ruled 681–742.[9] His son, Bird Jaguar IV acceded to the throne in 752, ten years after his father's death. Scholars have speculated about the reason for this interregnum, and many have concluded that Bird Jaguar IV's claim to the throne may have been threatened by other contenders or that he was not the rightful heir to the throne.[10] Questions of legitimacy of Maya rulers are difficult for scholars to

---

[9] Simon Martin and Nikolai Grube, *Chronicle of the Maya Kings and Queens: Deciphering the Dynasties of the Ancient Maya* (London: Thames and Hudson, 2000), p. 123.

[10] See Tatiana Proskouriakoff, 'Historical Data in the Inscriptions of Yaxchilan: Part I', *Estudios de Cultura Maya*, no. 3 (1963): pp. 149–67 for the first suggestion of this nature; see also Martin and Grube, *Chronicle of the Maya Kings and Queens*, p. 127; Sarah Bardsley, 'Rewriting History at Yaxchi-

understand, primarily because definitive evidence is lacking to state how the inheritance of kingship was maintained among the Classic Maya rulers, whether by primogeniture or not.[11] Compelling evidence suggests that authority and ritual privileges were instead transferred through ritual events and the passing of ceremonial objects.[12] It is possible that heir designation rituals were used to formerly support a particular claim to the throne, although, as Patrick Culbert suggests, complications in succession were considered normal.[13] Inheritance from father to son appears to have been common in the Classic period Maya rulership, although there is no way to determine whether heirs were the eldest sons or not, and there are a number of cases of women inheriting the throne, such as in Palenque and Naranjo, or brothers of previous kings taking control, such as in Copan and Yaxchilan.[14]

Upon his enthronement, Bird Jaguar IV undertook a prolific building campaign, and commissioned a large number of carved monuments, including stelae, lintels, and stairways, suggesting that the king was trying to affirm his legitimacy to a greater degree than other contemporary Maya rulers.[15] However, it should be remembered that building campaigns and monumental construction were also used to demonstrate the wealth of rulers, and their ability to command the labour necessary for such programmes, and thus were symbols of a ruler's political and financial might.[16] To be able to afford such constructions, and demand the labour needed, Bird Jaguar IV must have been in a relatively secure position upon taking the throne in 752. It should also be noted that his father, Shield Jaguar II, undertook a similar spate of

---

lan: Inaugural Art of Bird Jaguar IV', in V. M. Fields, ed., *Seventh Palenque Round Table, 1989* (San Francisco: Pre-Columbian Art Research Institute, 1994), pp. 87–94; Nikolai Grube, 'Observations on the Late Classic Interregnum at Yaxchilan', in W. M. Bray and L. Manzanilla, eds., *The Archaeology of Mesoamerica: Mexican and European Perspectives* (London: British Museum Press, 1998), pp. 116–27; Kathryn Josserand, 'The Missing Heir at Yaxchilan: Literary Analysis of a Maya Historical Puzzle', *Latin American Antiquity* 18, no. 3 (2007): pp. 295–312.

[11] Heather Irene McKillop, *The Ancient Maya: New Perspectives* (Santa Barbara: ABC-Clio, 2004), 179.

[12] Christophe Helmke, 'The Transferral and Inheritance of Ritual Privileges: A Classic Maya Case from Yaxchilan, Mexico', *Wayeb Notes* 35 (2010): pp. 1–14.

[13] Patrick T. Culbert, 'Maya Political History and Elite Interaction: A Summary View', In P. T. Culbert, ed., *Classic Maya Political History: Hieroglyphic and Archaeological Evidence* (Cambridge: Cambridge University Press, 1991), p. 332.

[14] See Martin and Grube, *Chronicle of the Maya Kings and Queens.*

[15] See Caroline Tate, *Yaxchilan: The Design of a Maya Ceremonial City*, Kindle edition, (Austin: The University of Texas Press, 1992), loc 3793.

[16] David Webster, 'Classic Maya Architecture: Implications and Comparisons', in S. D. Houston, ed., *Function and Meaning in Classic Maya Architecture: A Symposium at Dumbarton Oaks, 7th and 8th October 1994* (Washington, DC: Dumbarton Oaks, 1998), pp. 35–36.

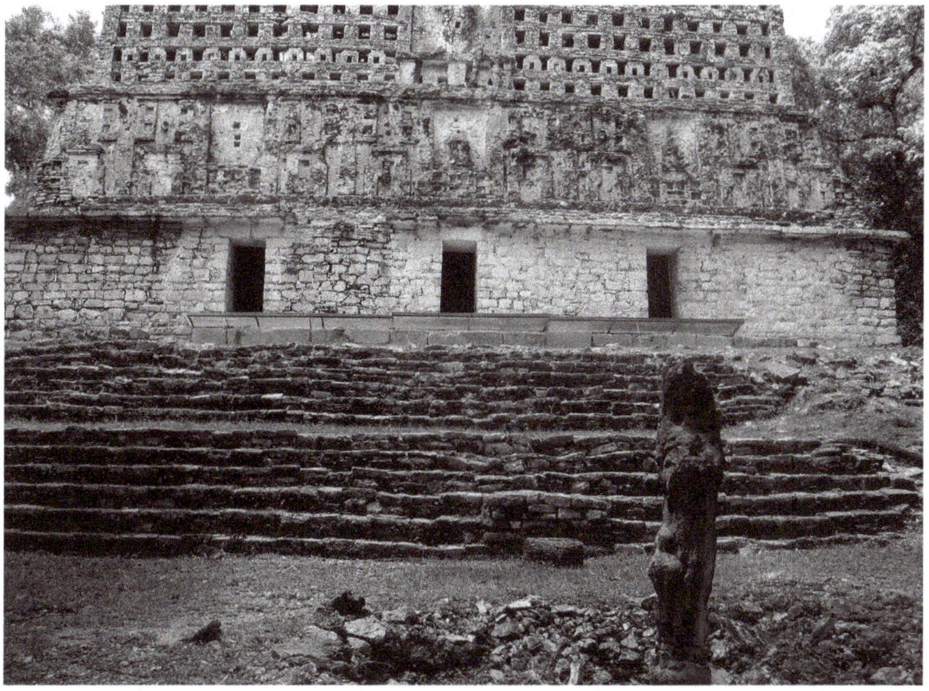

Figure 3.4: Structure 33, Yaxchilan, including placement of HS. 2. Step VII is the largest step of thirteen, and is in the centre. Author's photograph.

building activity during his reign, changing the built landscape of Yaxchilan completely, although this has never been attributed to an insecure reign.[17]

HS. 2 was commissioned along with its accompanying structure (Structure 33, see FIGURE 3.4) after Bird Jaguar IV's succession in 752, although the event alluded to in the historical section of the text on Step VII (I1–R7) occurred in 744. Bird Jaguar IV is shown mid-play, engaged in a hip ballgame ritual and fielding a prisoner-as-ball against a set of six inscribed steps. He is watched by two dwarfs who stand to his left side (although on the right side of the step), within a metaphorical cave constructed by hieroglyphics (see below for discussion; FIGURE 3.1). The date of the ritual (FIGURE 3.1, glyph blocks M1–P2) suggests that Bird Jaguar IV had already been passed the authority to perform such a ceremony before officially acceding to the throne in 752, but did not have the ability (perhaps wealth, influence, or privilege) to commission the monument at that time. There are a number of theories as to why this is, which I will not elaborate on here, but to say that there is little consensus among

---

[17] Tate, *Yaxchilan: The Design of a Maya Ceremonial City*, loc 3722–47.

scholars.[18] Bird Jaguar IV is depicted playing the ballgame with an audience of two dwarfs representing deities and celestial bodies, perhaps because he wished to illustrate the celestial sanction for his rule before his official enthronement.

HS. 2 was part of a larger ceremonial space surrounding Structure 33. The steps line the uppermost riser of thirteen, leading into the building. Thirteen was a significant number to the ancient Maya: it was the number of layers in the celestial realm; it was connected with a specific deity, the Water Lily Monster, who was associated with the earth, sky, and watery Underworld; it was associated with rain, water, and blood[19]; and it was mathematically important in the construction of the Long Count and Calendar Rounds of the Maya calendar.[20] The ceremonial space of Structure 33 was deliberately designed to evoke the different layers of the Maya universe, and create a 'cosmic map'.[21] Bird Jaguar IV used this space to support his ordained right to rule over Yaxchilan, and reaffirm his divine power within the earthly realm and beyond. HS. 2, and the dwarfs that appear there, are part of this larger message.

Dwarfs in Classic Maya Imagery and Ideology

Dwarfs appear in the imagery of the ancient Maya Lowlands and beyond, across time, although scholars should be reluctant about making ethnographic analogy between the Maya and other Central American groups' treatment of them. As Virginia Miller explains, this is because there are no accounts in contemporary literature of dwarfs in Maya culture from the conquest era or before.[22] Classic period Maya depicted dwarfs, hunchbacks, and other deformed humans in figurines, polychrome vases, and on monumental sculpture, as well as in jade plaques and other worn decorative items.

---

[18] Instead see Martin and Grube, *Chronicle of the Maya Kings and Queens*; Bardsley, 'Rewriting History at Yaxchilan'; Nikolai Grube, 'Observations on the Late Classic Interregnum at Yaxchilan'; Josserand, 'The Missing Heir at Yaxchilan'; and Tate, *Yaxchilan*, among others.

[19] Merle Green Robertson, 'The Celestial God of Number 13', *The PARI Journal* 12, no. 1 (2011): pp. 1–6.

[20] See Sven Gronemeyer and Barbara MacLeod, 'What Could Happen in 2012: A Re-analysis of the 13-*Bak'tun* Prophecy on Tortuguero Monument 6', *Wayeb Notes* 34 (2010), and Marcel Polte, 'Significance of the Coefficient 13 in the Long Count Calendar', *Wayeb Notes* 41 (2012).

[21] Suzanne Nolan, 'Late Classic Politics and Ideology: A Case Study of Hieroglyphic Stairway 2 at Yaxchilan, Chiapas, Mexico' (PhD, University of Essex, 2015)

[22] Virginia Miller, 'The Dwarf Motif in Classic Maya Art' In M. G. Robertson and E. P. Benson eds. *Fourth Palenque Round Table, 1990* (San Francisco: Pre-Columbian Art Research Institute, 1985), p. 142.

Commonly, Maya artists and sculptors depicted dwarfs with the following characteristics: small stature, large disproportioned head and prominent brow, short fleshy limbs, pot-belly, and dropped lower lip.[23] These features in artistic representation have led scholars to identify that the majority of dwarfs in Maya art probably had achondroplasia, a common cause of dwarfism that accounts for 80% of cases.[24] The condition is genetic, and achondroplasia is a result of a mutation on chromosome 4. Representation in Maya imagery suggests that this was a well known condition within the culture, although it is probable that depictions overemphasised the number of living dwarfs within the society, in which case figurines and other images were used as substitutes. Currently, there is little archaeological evidence of dwarfs in the Maya lowlands, although it is possible that in Burial 24 at Tikal a dwarf was buried with a ruler.[25]

That dwarfs were overrepresented within Maya imagery suggests they were highly valued among the elite community, and their presence – whether real or figurative – was sought after. Dwarf representations alongside rulers and gods have led scholars to suggest that some were indeed historical individuals that may have enjoyed an elevated and unique position with the Maya royal courts.[26] Dwarfs would provide entertainment for the nobility through dance, serve food to the ruling elite, and had a wide range of administrative duties including taste testing and assessing the quality of tribute (see K1453 for an example, FIGURE 3.5).[27]

Many dwarfs acted as courtiers to Maya rulers, and held important roles alongside the ruling elite, such as those outline above. They also appeared frequently alongside deities, in particular the Maize God and the Sun God of the Maya (such as on K517, FIGURE 3.6, and K8533, FIGURE 3.7).[28] Christian Prager argues that rulers

---

[23] Miller, 'The Dwarf Motif in Classic Maya Art', 141.

[24] See Miller, 'The Dwarf Motif in Classic Maya Art'; also Christian M. Prager, 'Court Dwarfs: The Companions of Rulers and Envoys of the Underworld', in Nikolai Grube ed. *Maya: Divine Kings of the Rainforest* (Cologne: Könemann, 2006), pp. 278–79; and Takeshi Inomata, 'The King's People: Classic Maya Courtiers in a Comparative Perspective', in S. D. Houston and T. Inomata eds. *Royal Courts Of The Ancient Maya: Volume 1: Theory, Comparison, And Synthesis* (Boulder, CO: Westview Press, 2000), pp. 27–53.

25 William R. Coe, Tikal Report 14: Excavations in the Great Plaza, North Terrace, and North Acropolis of Tikal. University Museum Monograph 61 (Philadelphia: University of Pennsylvania, 1990), pp. 540–43.

[26] Inomata, 'The King's People', pp. 36–37.

[27] Phil Wanyerka, 'A Fresh Look at a Maya Masterpiece' *Cleveland Studies in the History of Art* 1 (1996): 81; Prager, 'Court Dwarfs'.

[28] Mary Miller and Karl Taube, *An Illustrated Dictionary of the Gods and Symbols of Ancient Mexico and the Maya* (London: Thames and Hudson, 1997), p. 82; Prager, 'Court Dwarfs'; Judith Storniolo,

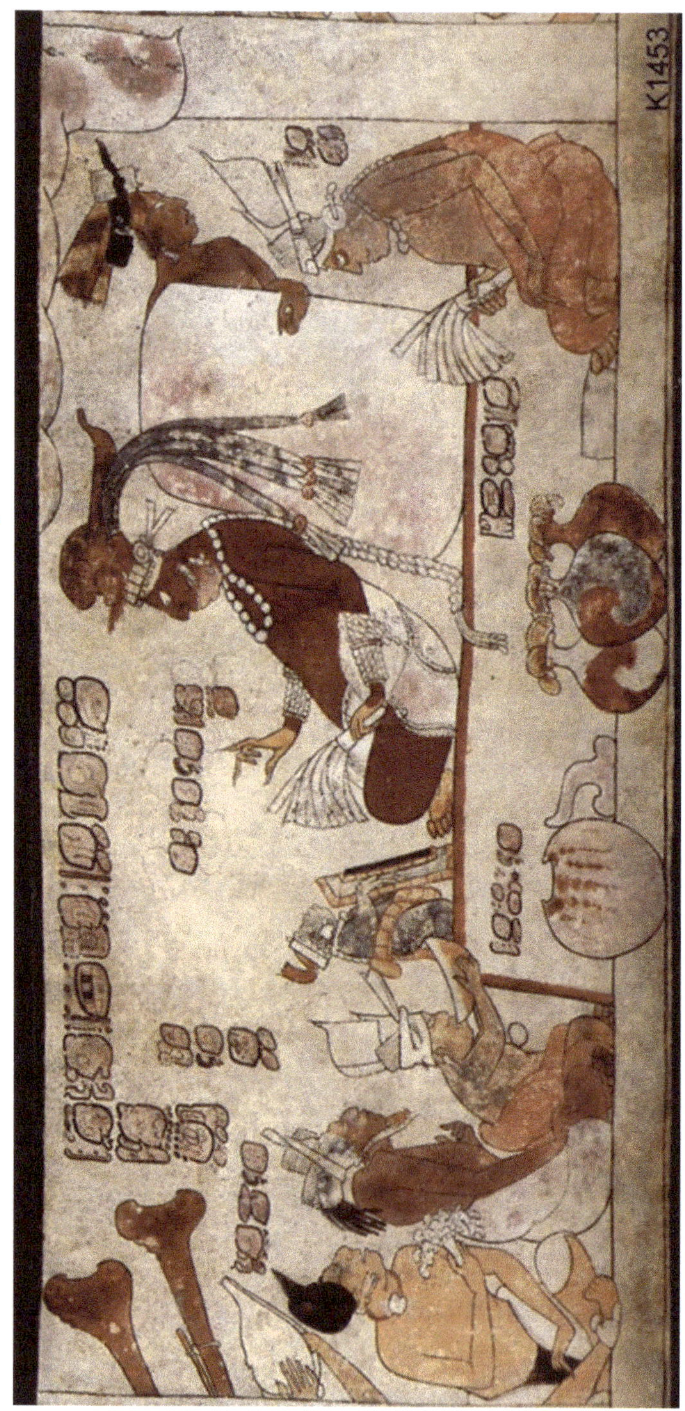

Figure 3.5: K1453, showing a palace scene with two dwarfs. One is holding an obsidian mirror for the ruler, the other is drinking from a bowl (perhaps as a taste-tester). Photographs © Justin Kerr, reproduced with kind permission from J. Kerr.

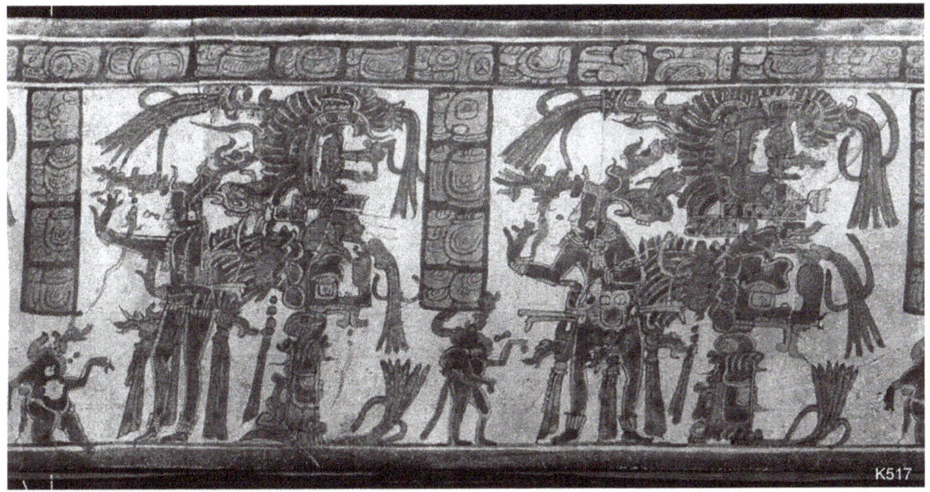

FIGURE 3.6: K517, showing the Maize God dancing with a dwarf. Note the earspools. Photographs © Justin Kerr, reproduced with kind permission from J. Kerr.

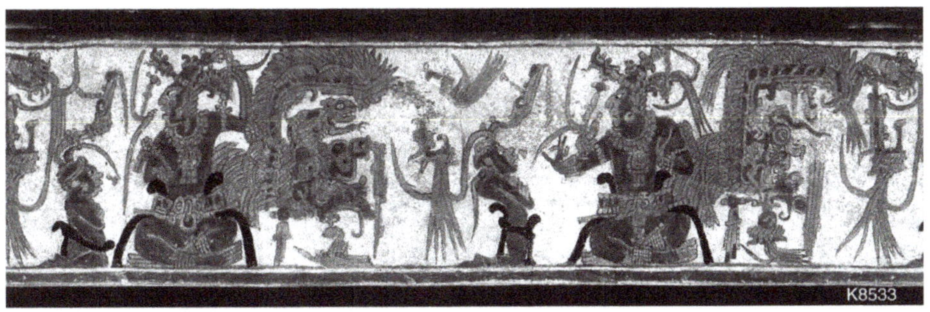

FIGURE 3.7: K8533, showing the Maize God sitting, and dwarf kneeling. Note the water bird headdress worn by the dwarf. Photographs © Justin Kerr, reproduced with kind permission from J. Kerr.

accompanied by dwarfs in rituals could take on the roles of those deities, and that the dwarfs' presence was part of the overall message of impersonation and visual metaphor.[29] Maya rulers would often wear the attire of a deity, or represent themselves alongside visual markers such as dwarfs, to indicate the impersonation (although not transformation) of particular gods.[30] At El Peru, for example, Na Kan

---

'Ancient Mythical Dwarfs in Modern Yucatan', *Expedition* 51, no. 1 (2009): p. 24. See also: K633, K1837, K3388, K3400, K4619.

[29] Prager, 'Court Dwarfs', p. 279.

[30] Andrea Stone, 'Aspects of Impersonation in Classic Maya Art', in M. G. Robertson and V. M.

Ajaw (a royal woman) is dressed as the Maize God and accompanied by a richly adorned dwarf.[31]

Dwarfs as companions to the Maize God, or in contexts that allude to the Maize God, suggest that the Maya believed they were closely linked to fertility and sustenance. Mary Miller and Karl Taube have suggested that the connection between the Maize God and dwarfs mirrored the presence of the smaller ear of maize often issued by a maize plant.[32] Dwarfs are connected to fertility throughout Mesoamerica. In Olmec imagery, dwarfs are often seen carrying ears of corn and bearing maize iconography.[33] Taube notes that an Olmec dwarf found in the Dumbarton Oaks collection may have been a symbolic bringer of maize to humanity.[34]

The Maya also believed that dwarfs were the offspring of the four aspects of Chac, the Maya god of rain and lightning.[35] Because the ancient Maya relied on the annual rains to water their crops, Chac was closely associated with the Maize God as a bringer of sustenance. Chac's importance has endured throughout history, and in contemporary Maya mythology he is responsible for breaking open a great rock that contained maize.[36] In ancient Maya belief, Chac took on the role of 'splitter', splitting open the mountain which contained maize, bringing the people sustenance.[37] He is often seen wielding an obsidian axe, a metaphor for lightning, and uses his weapon to decapitate his victims. Chac is also closely associated with serpents, creatures which are able to permeate every layer of the cosmos – the celestial, terrestrial, and Underworld. As Stone and Zender explain '[Chac's] dual personality, part macho aggression, part fertility bearer, made him a superb model for Classic Maya kings.'[38]

Dwarfs are not only connected to Chac, but are associated with rain and lightning in their own right. 'Rain dwarfs' acted as rain spirits in ancient Mexican beliefs, and were responsible for lightning.[39] While scholars must bear Miller's warnings in mind, there are examples of dwarfs bearing water symbols in Classic period Maya

Fields, eds., *Sixth Palenque Round Table, 1986* (Norman: University of Oklahoma Press, 1991), pp. 194–202.

[31] Wanyerka, 'A Fresh Look at a Maya Masterpiece', figure 1, p. 73.

[32] Miller and Taube, *Gods and Symbols of Ancient Mexico and the Maya*, p. 82.

[33] Peter David Joralemon, *A Study of Olmec Iconography* (Washington, DC: Dumbarton Oaks, 1971), figure 20f, p. 52.

[34] Karl A. Taube, *Olmec Art at Dumbarton Oaks* (Washington, DC: Dumbarton Oaks, 2004), p. 82.

[35] Miller and Taube, *Gods and Symbols of Ancient Mexico and the Maya*, p. 82.

[36] Miller and Taube, *Gods and Symbols of Ancient Mexico and the Maya*, p. 60.

[37] Stone and Zender, *Reading Maya Art*, p. 41.

[38] Stone and Zender, *Reading Maya Art*, p. 41.

[39] John E. Staller and Brian Stross, *Lightning in the Andes and Mesoamerica: Pre-Columbian, Colonial, and Contemporary Perspective* (Oxford: Oxford University Press, 2013), p. 143.

art, such as a Copan ballcourt marker showing a ballplayer wearing a 'bird-headed 'rain dwarf' as a frontal adornment.'[40] This example also demonstrates that the 'rain dwarf' may have had some association with the hip ballgame played by the ancient Maya at Copan (and Yaxchilan as seen by their presence on HS. 2). John Staller and Brian Stoss have also identified references to a White Dwarf and a Red Dwarf as Maya versions of the Central Mexican rain dwarfs.[41] Rain dwarfs were believed to have lived in caves, important transitional spaces in Maya ideology between the surface of the earth and the Underworld and sacred mountain spaces.[42]

Caves were also closely associated with the Mountain of Sustenance, where maize was first grown, and where Xmucane, mother to the Maize God before he was resurrected, ground the maize to make the first humans.[43] Caves were thus locations of a twofold nature: dangerous and dark, where the Underworld could be accessed through ritual and sacrifice, and connected with fertility and life. These kinds of dualities were common within Maya ideology, whereby supernatural creatures and places represented both positive and negative aspects.

The dwarfs on HS. 2 at Yaxchilan

Two dwarfs accompany the ruler Bird Jaguar IV on Step VII of HS. 2. They stand on his left side watching the king as he plays a hip ballgame against a set of steps using a prisoner-as-ball. These two entities probably represent mythological or celestial beings, rather than historical figures, and while they are dressed simply, they each wear some significant items that serve to identify their supernatural natures, in particular: their headdresses, earspools, and the glyphs under their arms. These features of their representation help to identify some of their cultural associations at Yaxchilan, and thus what they represent in this context. It is important to remember that while we can make certain assumptions around this, we – as scholars of the ancient Maya – will never be able to truly empathise with or understand the whole message, given our removal from the historical and ideological context in which these images were created.

---

[40] H. E. M. Braakhuis, 'The Tonsured Maize God and Chicome-Xochitl as Maize Bringers and Culture Heroes: A Gulf Coast Perspective', *Wayeb Notes* 32 (2009): p. 23.

[41] Stalle and Stross, *Lightning in the Andes and Mesoamerica*, p. 143.

[42] Jesper Nielsen and Christophe Helmke, 'Reinterpreting the Plaza de Los Glifos, La Ventilla, Teotihuacan' *Ancient Mesoamerica* 22, no. 2 (2011): p. 358.

[43] Dennis Tedlock, *Popol Vuh: The Definitive Edition of the Maya Book of the Dawn of Life and the Glories of Gods and Kings* (New York: Touchstone, 1996), pp. 139–40.

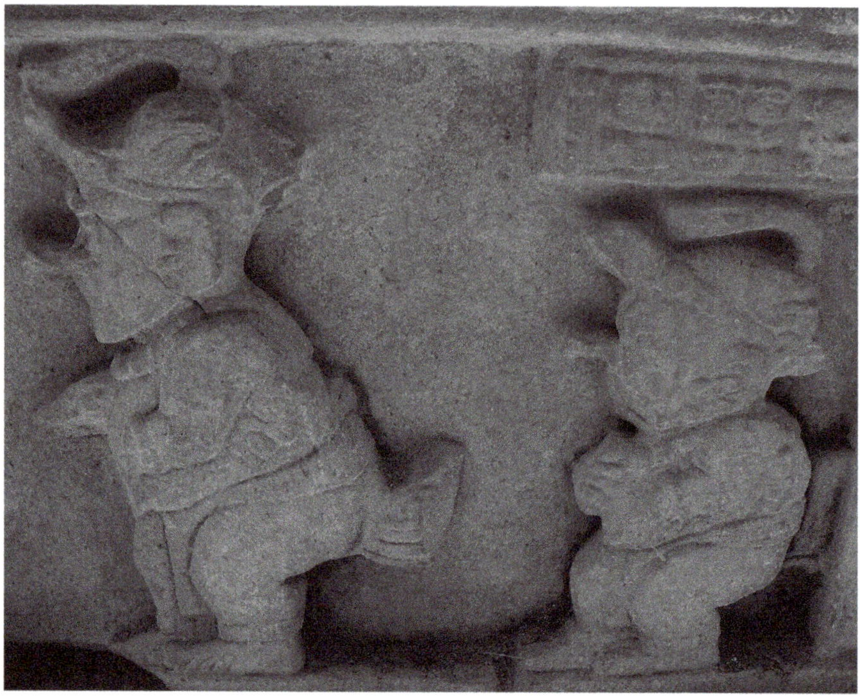

Figure 3.8: Detail of the two dwarfs on HS. 2, Step VII. Author's photograph.

Despite high levels of erosion, it is possible to identify some of the key features of the headdresses that the two dwarfs on HS. 2 wear. From figures 3.1 and 3.8, one can see that the dwarfs are wearing an animal head attached to their headdresses that have elongated necks and facial features. Because of the level of degradation on the original step, these creatures look almost serpent-like. However, there are a number of other examples of dwarfs wearing such headdresses, such as on K517 (FIGURE 3.6) and on K8533 (FIGURE 3.7). The latter provides the clearest example, with the head-dresses clearly showing cormorants or water bird heads.[44] There are no examples (to this author's knowledge) of dwarfs wearing serpent headdresses in Maya art, thus it is probable that on HS. 2 the dwarf figures are wearing cormorant headdresses.

The cormorant appears frequently on ancient Maya polychrome pottery alongside the Maize God or the Hero Twins (for example on K517: FIGURE 3.6, K626, K1004, and K5833: FIGURE 3.7). The Hero Twins were instrumental in the resurrection of the Maize God. During their adventures in the Underworld, they were burned and

____

[44] Braakhuis, 'The Tonsured Maize God and Chicome-Xochitl as Maize Bringers and Culture Heroes', p. 32.

their ashes cast into a river where they were reborn as catfish. It is possible that they were plucked from the water by cormorants,[45] whereupon they transformed into vagabonds and eventually defeated the lords of the Underworld. Water birds, such as cormorants, may have been used in Maya art to evoke transition, as they were able to cross between the different layers of the cosmos: in flight across the heavens, the terrestrial, and the watery Underworld while hunting. On Stela 14 at Dos Pilas, the ruler is accompanied by a dwarf and a cormorant with a fish in its beak. Matthew Looper suggests that this recalls dances performed by the Maize God on polychrome ceramics from the Naranjo-Holmul region, once again affirming the connection between dwarfs and the Maize God.[46]

The headdresses worn by the dwarfs on HS. 2, then, recall the Maize God, fertility, and the transition between the different layers of the Maya universe. The use of the cormorant headdresses may represent a conflation of the imagery used at Dos Pilas, on Stela 14, where the dwarf and cormorant are depicted separately (conflation being a common artistic technique in Classic Maya art[47]). As discussed above, dwarfs often accompanied the Maize God and Maya rulers impersonating the Maize God. It is possible that culturally embedded understanding of the connection between the Maize God and dwarfs meant that the dwarfs on HS. 2 were used as an ideological connection with maize and fertility, as well as the watery Underworld.

Moving on to analysis of the earspools the dwarfs on HS. 2 wear, it is clear that the Maize God associations continue. The smaller of the two dwarfs wears an earspool of a design that is often worn by the young Maize God on polychrome ceramics, such as on K517 (FIGURE 3.6), K1837 and K4619. Earspools are closely connected with wind and breath, and may have been containers for life-essences.[48] It is possible, therefore, that by wearing the earspool of a particular deity, the dwarfs on HS. 2 were not just representatives of that deity, but literally carried a piece of their life-essence with them, so that the deity in question was present in spirit to watch the ballgame and its outcome.

---

[45] Justin Kerr, 'Kerr Number: 6609' at http://research.mayavase.com/maya_selects.php?image_number=6609undefined [accessed 12/01/2015]; also see Michael J. Grofe, 'Recipe for Rebirth: Cacao as Fish in the Mythology and Symbolism of the Ancient Maya' (PhD Dissertation, University of California, 2007)

[46] Looper, *To Be Like Gods*, p. 25.

[47] For explanation on conflation see Stone and Zender, *Reading Maya Art*, p. 21.

[48] Karl A. Taube, 'The Symbolism of Jade in Classic Maya Religion', *Ancient Mesoamerica* 16 (2005): pp. 23–50; Karon Winzez, 'The Symbolic Vocabulary of Cloth and Garments in the San Bartolo Murals', in Heather Orr and Matthew G. Looper, eds., *Wearing Culture* (Boulder, CO: University Press of Colorado, 2014), pp. 373–410.

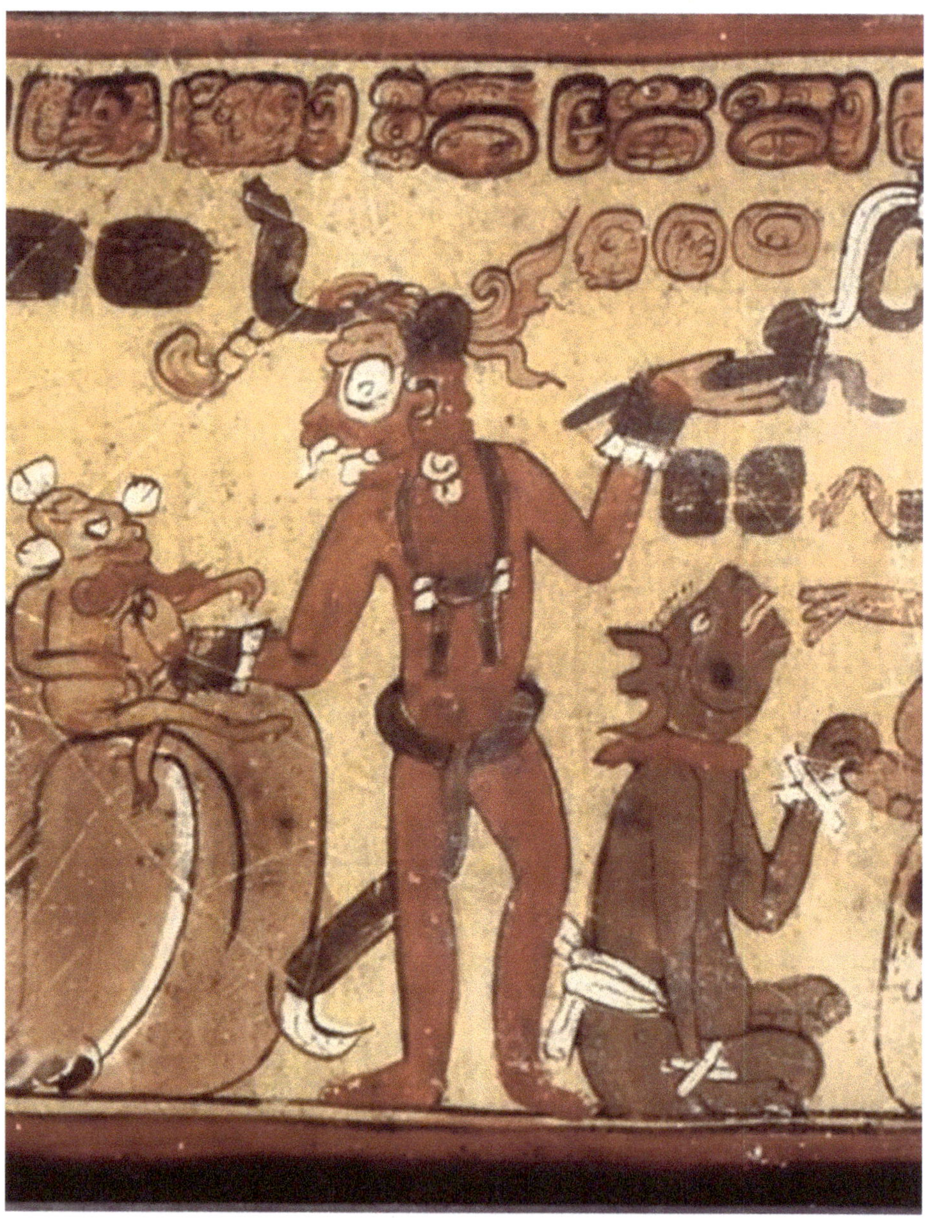

Figure 3.9: Detail of K555, showing a scene with Chac. Note the spondylus shell earspool. Photographs © Justin Kerr, reproduced with kind permission from J. Kerr.

The larger of the two dwarfs wears a spondylus shell earspool that is often seen being worn by the god Chac. The rain god is frequently seen wearing this design, including in his name glyph,[49] and on polychrome ceramics (see K555, FIGURE 3.9, as well as K521 and K1152 among others). As the god of rain, Chac is connected to fertility and the Maize God, as discussed above, further solidifying those associations of the dwarfs.

The dwarfs on HS. 2 may have also been a visual reference to the story written on the left side of the carved step (FIGURE 3.1). In the story, carved onto the steps against which Bird Jaguar IV is playing the ballgame (glyph blocks A1-H6), three aspects of the Maize God (seen on glyph blocks B2, C2, and E4-F4) are sacrificed in a *ch'ak-ab'*, or beheading, ritual (seen on glyph blocks A2, C1, and E3).[50] As outlined above, Chac's weapon of choice was an obsidian axe, and he was responsible for splitting the mountain to free the maize for humans to eat. It is possible that the *ch'akab'* ritual involved Chac in some way, and that the dwarfs on the right side of the step were used to mirror this mytho-historical story, whereby the beheading of three aspects of the Maize God brought sustenance to the Maya people. It should be noted that two of the three aspects of the Maize God (C2 and E4-F4) have not been discovered on any other Maya text to date, and may have been patron deities at Yaxchilan alone (and perhaps even only during the Late Classic period).[51]

The final components to be discussed can be found under the upper arms of both dwarfs. The 'm' shape under the arms are embedded *ek'* glyphs, which are generally taken to mean 'star'. This is most certainly a glyphic indicator, used to designate the dwarfs as particular types of creature, or possessing particular qualities. In context, however, the meaning is difficult to decipher. Caroline Tate suggests that the glyph connects the two dwarfs with Jupiter's appearance in the night sky on the last calendar round date inscribed upon the monument, when Bird Jaguar IV played the ballgame depicted.[52] On this date, Jupiter passed by Castor and Pollux (the two brightest stars in Gemini), which Tate suggests could be represented by the dwarfs.[53] Prager is more cautious, and suggests that the two dwarfs on HS. 2 could represent a constellation that has yet to be understood by scholars.[54] Other stars and constel-

---

[49] Michael D. Coe and Mark van Stone, *Reading Maya Glyphs* (London: Thames and Hudson, 2001), p. 111.

[50] Prager, 'Court Dwarfs', pp. 278–79.

[51] For further discussion, see Nolan, 'Late Classic Politics and Ideology', pp. 139–49.

[52] Tate, *Yaxchilan: The Design of a Maya Ceremonial City*, location 4038.

[53] Tate, *Yaxchilan*, location 4038; see also Susan Milbrath, *Star Gods of the Maya: Astronomy in Art, Folklore, and Calendars* (Austin: University of Texas Press, 2000), p. 268.

[54] Prager, 'Court Dwarfs', p. 279.

lations have been identified within Maya hieroglyphic texts: Orion is *ahk ek'* and Scorpio is *sinan ek'*.[55] However, neither of the *ek'* glyphs attached to the dwarfs on HS. 2 come with additional glyphic markers, making it difficult to determine which star or constellation they may refer to. Venus is often referred to as *chak ek'* in Maya hieroglyphic texts, suggesting that the larger of the two dwarfs (with the features connecting it with the rain god *Chac*) may represent this planet. This is, however, highly tentative, and relies heavily on the interchangeability of use of the glyph *chak* for 'red/great', and the use of the word Chac for the deity.[56] Similarly, it is possible that the dwarfs are connected with the so-called 'turtle constellation', Orion, because of the homophony between *aak*, meaning 'dwarf', and *ahk*, for 'Orion'.[57]

When appearing on its own, *ek'* could also represent one of the most important celestial bodies in ancient Maya ideology, Venus. Michael Closs has argued that 'the *ek'* glyph without prefixes is found in celestial bands together with symbols of the sun and moon', suggesting that we should accept that *ek'* is a symbol for Venus, as the sun, the moon and Venus all play a highly significant role in Maya ideology.[58] The *ek'* glyph as representative of Venus appears at a number of other archaeological sites across the Maya world. The Palace of the Governor at Uxmal, for example, is decorated with sky bands of sun, moon, and *ek'* glyph motifs and its orientation is directly related to the planet Venus.[59] Venus was a principle celestial body for the ancient Maya, and it was closely associated with rain, maize,[60] and warfare and military strength.[61] Ivan Šprajc has also discussed the Venus-Rain-Maize complex in great detail,[62] suggesting a clear connection between the *ek'* glyph (as representative of Venus), Chac (as a representative of rain), and the Maize God.

---

[55] Stone and Zender, *Reading Maya Art*, p. 61.

[56] For examples of where homophony is used within Maya hieroglyphic script, see Stephen D. Houston, 'An Example of Homophony in Maya Script', *American Antiquity* 49, no. 4 (1984): pp. 790–805.

[57] Milbrath, *Star Gods of the Maya*, pp. 266–69; Prager, 'Court Dwarfs', p. 279.

[58] Michael Closs, 'Venus in the Maya World: Glyphs, Gods, and Associated Astronomical Phenomena', in M. Green Robertson and D. Call Jeffers, eds., *Tercera Mesa Redonda de Palenque*, Vol. IV (Monterey, CA: Pre-Columbian Art Research, 1979), p. 148.

[59] Ivan Šprajc, 'Governor's Palace at Uxmal'. In C. L. N. Ruggles, ed. *Handbook of Archaeoastronomy and Ethnoastronomy* (New York: Springer, 2015), p. 779.

[60] See Ivan Šprajc, 'The Venus-rain-maize complex in the Mesoamerican world view: part I', *Journal for the History of Astronomy* 24 (Parts 1/2) (1993a): pp. 17–70, and Šprajc, 'The Venus-rain-maize complex in the Mesoamerican world view: part II'. *Archaeoastronomy* 18 (*Journal for the History of Astronomy*, Supplement to Vol. 24) (1993b): pp. S27–S53; Storniolo, 'Ancient Mythical Dwarfs', 22.

[61] Foster, *Handbook to Life in the Ancient Maya World*, p. 262.

[62] Šprajc, 'The Venus-rain-maize complex in the Mesoamerican world view: part I'; Šprajc, 'The Venus-rain-maize complex in the Mesoamerican world view: part II'.

What is clear is that the dwarfs were being associated with the stars, and thus the heavens. The glyphs for stars marked them out as connected with celestial bodies. This was used to contrast their placement within the composition of the step, where they stand off to one side, surrounding along the top and to their backs with glyphs, which is a deliberate allusion to a cave.[63] Caves were (and are) of the upmost importance within Maya ideology, being liminal spaces between the earth's surface and the Underworld.[64] As discussed above, they were locations of both danger and fertility, where communication between kings and gods could take place. By depicting the dwarfs on HS. 2 within a cave space, the artist may have been demonstrating that while the dwarfs are observing the actions of Bird Jaguar IV, they are not of the ballgame themselves. They stand not as participants, but as passive watchers and representatives of the gods of which they bear the markings.

The combination of the *ek'* glyphic markers, the water bird headdresses, earspools recalling the Maize God and Chac, and Venus, and the dwarfs presence within the cave combine to indicate that the dwarfs on HS. 2 are creatures that are connected to every layer of the Maya universe. They are at once a metaphor for the stars (*ek'*), the earth (watching the ballgame), and the Underworld (in their placement in the cave), while also denoting the presence of the Maize God and Chac. Bird Jaguar IV used the associations (and, no doubt, others) to demonstrate that the ballgame that he plays on Step VII, and by extension his reign over Yaxchilan, was not only sanctioned by the gods, but was supported at every layer of the cosmos.

Recalling that there was a ten year gap between his father's death and Bird Jaguar IV's accession, it is probable that the use of the layered visual metaphors within the representation of the dwarfs on HS. 2 (among other artistic choices on the hieroglyphic stairway, and across the monuments he commissioned) were used to reaffirm his legitimacy to rule.

Conclusions

This chapter has explored the metaphorical associations of dwarfs in Classic period Maya imagery. It has summarised their activities within terrestrial court life, and discussed the role they played as supernatural agents within ancient Maya ideology. Dwarfs accompanied both rulers and gods, acting as assistants to daily tasks and important rituals that served to keep the cosmos in balance.

---

[63] Storniolo, 'Ancient Mythical Dwarfs', p. 22.

[64] Evon Z Vogt and David Stuart, 'Some Notes on Ritual Caves among the Ancient and Modern Maya', in J. E. Brady and K. M. Prufer, eds., *In the Maw of the Earth Monster: Mesoamerican Ritual Cave Use* (Austin: University of Texas Press, 2005), p. 179.

On HS. 2, at Yaxchilan, the presence of the dwarfs served a number of functions in the support of the ruler Bird Jaguar IV. Their attendance as watchers of the ballgame played by Bird Jaguar IV gave the ritual divine approval. Furthermore, their presence suggested to onlookers that the authority with which Bird Jaguar IV's ruled Yaxchilan was supported by the Maize God and Chac, whom the dwarfs' also represented. The two gods were ideal models for Maya rulers on which to base their authority, given their associations with both strength and fertility. The possible associations with Venus support this, along with the allusion that Bird Jaguar IV was a militarily strong ruler.

The headdresses worn by the dwarfs, and contrast of the *ek'* glyphs and placement within the cave constructed from hieroglyphs, suggest that they were supernatural creatures which could touch and influence every layer of the Maya universe. Water birds access each layer of the cosmos, and it is possible that their presence alluded to the dwarfs' ability to emulate this. Placement within the cave further suggests a connection to the Underworld, as well as fertility, and perhaps further recalls the Maya belief that maize originated within a cave, and was 'broken free' by the god Chac. Thus, the dwarfs were metaphors for stars, the earth, and the Underworld, as well as sustenance, and strength.

For Bird Jaguar IV, this meant that the ballgame ritual and his rulership was supported at each layer of the Maya universe, and that he was able to call on supernatural supporters to attend and watch his most important rituals. Through this visual composition, he was also connecting himself to each layer of the cosmos. Finally, the dwarfs alone served as a method for reinforcing his right and capacity to rule over Yaxchilan, at a time when his legitimacy may have been in question.

ASTROLOGICAL IMAGERY IN THE RULERSHIP PROPAGANDA OF DUKE COSIMO I DE' MEDICI: THE VILLA CASTELLO

Claudia Rousseau

ABSTRACT: As I have previously demonstrated, the impresa depicting the Sign of Capricorn with the constellation of the Crown of Ariadne invented soon after the 18 year-old Cosimo de' Medici was confirmed second hereditary Duke of Florence in September 1537 was neither a vague reference to the Emperor Augustus, nor to Holy Roman Emperor, Charles V. Rather, the image that appeared on a medal commissioned immediately after this event was based on significant data from the youth's natal map. It became the core image of a rulership propaganda that would make continuous reference to a destiny indicated in the stars at his birth and was repeated in visual expressions throughout his career. Its astrological source and meaning were key to its effectiveness. In fall 1537 Cosimo commissioned painter Jacopo Pontormo to represent the positions of the full chart in the loggia vault of the Medici villa at Castello. Numerous works of art made for the Duke would include astrological imagery, including a sculpture by Domenico Poggini, another by Vincenzo Danti, and many examples in the decoration of the Palazzo Vecchio. Cosimo's use of an astrological basis for his rulership propaganda would become a model for later European monarchs such as Elizabeth I of England.

The astrological *impresa,* or personal emblem, that was invented soon after the eighteen-year-old Cosimo de' Medici was confirmed second hereditary Duke of Florence in September 1537, was not, as some scholars have suggested, a self-comparison to the Emperor Augustus, nor done in some imagined emulation of Charles V, the Holy Roman Emperor, who had, in effect, put him on the ducal throne.[1] More significantly, the *impresa* depicting the goat-fish Sign of Capricorn with the constellation of the Crown of Ariadne that first appeared on a medal (FIGURE 4.1) commissioned immediately after the confirmation, and was presumably widely circulated, was based on significant data from Cosimo's natal map. It became the core image of a rulership propaganda that would make continuous reference to a destiny

---

[1] My interpretation of the *impresa* first appeared in Claudia Rousseau, 'Cosimo I de' Medici and Astrology: The Symbolism of Prophecy', (PhD dissertation, Columbia University, 1983), Chapter I where its particular astrological basis is thoroughly demonstrated. Although Augustus did use Capricorn as a personal symbol, Charles V did not. For a summary of previous interpretations of the meaning of the *impresa,* see pp. 59–62 of that work. I have repeated my argument in 'The Pageant of Muses at the Medici Wedding of 1539 and the Decoration of the Salone del Cinquecento' in Barbara Wisch and Susan Munshower, eds., *'All the World's a Stage…,' Art and Pageantry in the Renaissance and Baroque, Part 2: Theatrical Spectacle and Spectacular Theater* (State College, PA: Pennsylvania State University Press, 1990), pp. 416–57; 'An Astrological Prognostication to Duke Cosimo I de' Medici of Florence', *Culture and Cosmos* 3, no. 2 (Autumn/Winter 1999): pp. 31–58.

foretold in the stars at Cosimo's birth, and was repeated in artistic expressions, sometimes in variant forms, from the outset of his career to its very end. Among the interesting changes this represented from previous rulership themes employed by the Medici, underlining of the ever renewing strength of the family dynasty, Cosimo's own propaganda was focused on the notion of a *personal* destiny fulfilled. This idea was promoted through increasing stress on the significance of the young Duke's *own* astrological fortune, as well as on certain parallels between his rise to power and that of the Roman Emperor Augustus.[2]

Cosimo's extensive use of an astrological *impresa* that specifically referred to his natal chart proclaimed – first and foremost against the loud voices of his many detractors – that his unexpected rise to power and position was not because he simply had had the 'good luck' to be a Medici when one was needed to fill the ducal throne. Instead, it indicated that the good fortune that brought this youth, from the cadet branch of the family, living in almost total obscurity in the Mugello with his widowed mother, had been preordained in the stars by nothing less than Divine Providence. This propaganda, largely generated in direct response to the political difficulties of his early years, would be the source of the emergence of an iconography of absolute sovereignty that was new in Florence. It would become a model for later monarchs, including Elizabeth I of England and later, Cosimo's descendant, Louis XIV of France.[3]

---

[2] This understanding of the changed character of Cosimo I's rulership imagery and propaganda, from emphasis on the family to focus on the Duke himself, is made clear in my dissertation. It is in contrast to the views presented by Janet Cox-Rearick in her book *Dynasty and Destiny in Medici Art* (Princeton: Princeton University Press, 1984), although she acknowledged and employed my interpretation of the impresa in that publication.

[3] Queen Elizabeth's well known adaptation of the figure of Astraea-Virgo was first extensively documented by Frances Yates in her ground-breaking article 'Queen Elizabeth as Astraea', *Journal of the Warburg and Courtauld Institutes* 10 (1947): pp. 27–82. She elaborated the theme in her later book, Astraea, *The Imperial Theme in the Sixteenth Century* (London-Boston: Routledge & Kegan Paul, 1975). However, Yates never actually associated it with the fact that Elizabeth was born with the Sun at 24° Virgo and the ASC at 6° Capricorn that would have been for her the 'true' basis for this extensive propaganda. Ellizabeth's birth data was recorded in Hall's Chronicle as being 'vii day of September being Sondaie, between thre and foure of the Clocke after noone' (ed. 1809, p. 805). Other sources put the birth at 2:54 or 3 PM on the same day (cf. Sue Toohey, 'Elizabeth I, The Virgin Queen'. http://www.skyscript.co.uk/qe1.html [accessed 8 Jan 2017]). Louis XIV was also born with the Sun in Virgo well positioned in the 10th house. Apart from the imperial and golden age references that were possible from this fact, it was believed that Virgo was the 'horoscope [ASC] of France'. Cf. A. Favyn, *Theatre d'honneur et de chevalerie* (Paris: Robert Fouët, 1620), p. 275. Louis XIV was Cosimo's great/great grandson through the marriage of Catherine de' Medici to Henry II.

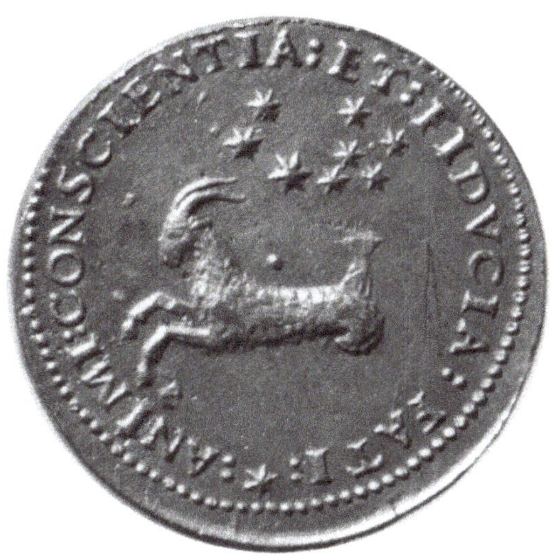

FIGURE 4.1: Domenico dei Vetri, Medal for Cosimo I de' Medici. Reverse: Capricorn/Crown of Ariadne/ANIMA CONSCIENTIA ET FIDUCIA FATI. 1537. Museo Nazionale del Bargello, Florence.

The previous decade had been a chaotic and violent one in Italy. Taking advantage of the horrific Sack of Rome in May 1527, including the captivity of the Medici Pope in Castel' Sant'Angelo, republican supporters in Florence had successfully expelled the Medici from the city. A deal negotiated between Pope Clement VII (Giulio de' Medici) and Emperor Charles V in June 1529 re-establishing them in Florence was not enforced without a grueling eleven-month siege of the city by Imperial forces. Soon after, Giulio's illegitimate son, Alessandro de' Medici, was confirmed first hereditary Duke of Florence in April 1532. Hated by many, he was murdered in January 1537 by his cousin, Lorenzino (Lorenzo di Pierfrancesco de' Medici). The accords between the Pope and the Emperor stipulated that in the event of Alessandro's death, and in the absence of legitimate male issue, the nearest legitimate male relative should succeed him. The reformed constitution of 1532 had repeated these stipulations. Cosimo was the only legitimate Medici male available, and with his youth and 'country air' (*quell' ragazotto dall'aria un po' campagnola*) the Florentine Senate hoped to be able to control him.[4] They, like the Florentine rebels (*i fuorusciti*) who had retreated to Siena where they were planning an offensive against him, miscalculated both his determination and strategic intellect, as well as his ability to turn the tides that were

---

[4] Benedetto Varchi, *Storia fiorentina*, Lelio Arbib, ed., (Florence: Società ed. delle storie del Nardi e del Varchi, 1838–1841), III, pp. 254–78; Rousseau, *Cosimo I*, pp. 35–36.

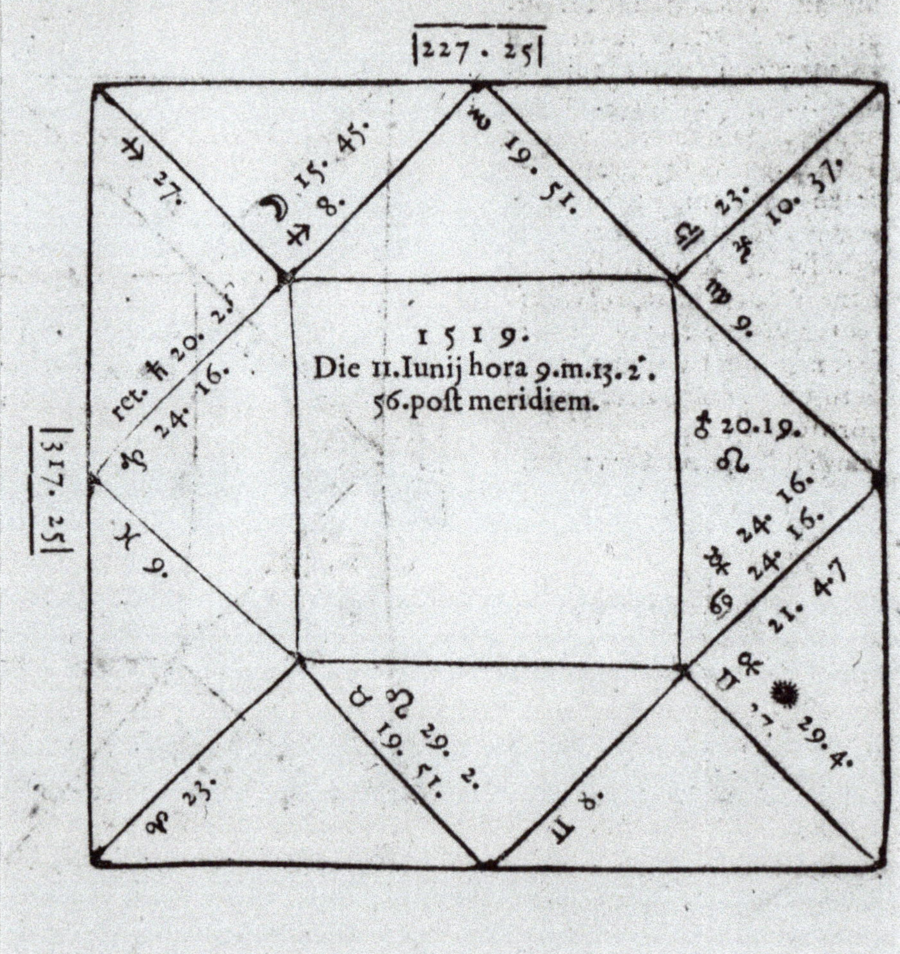

FIGURE 4.2: Francesco Giuntini, Nativity of Cosimo I de' Medici from *Speculum Astrologiae* (Lyons, 1581), vol. I, p. 127.

against him by using astrological and artistic tools that were particularly effective in the Florentine context.

There was, above all, an urgent need to contradict the prophecies of the *piagnoni* faction of the Dominicans and prove them false. These latter-day followers of Savonarola continued their provocations, even after the dramatic defeat of the *fuorusciti* at Montemurlo on 1 August 1537. They claimed, for example, that Alessandro's murder proved that God did not want a Medici to rule Florence, and that Cosimo would therefore, by the will of God, fail. In response, the Duke's rulership propaganda took on an increasingly messianic quality. It implied that it was God's Divine Providence that had sent Cosimo to settle the unrest and civil war that had plagued Florence, and to bring peace and prosperity to her people – the evidence for which was to be found in the stars at his birth. In the latter aspect Cosimo was indeed compared, repeatedly, to Octavian – Augustus – , on the basis of parallels believed to exist in their lives *because* of supposed parallels between their natal charts, and to a lesser extent, also to that of Charles V.

At this point we should turn to the basis of the astrological meaning of the Capricorn and Crown image that explains why it was related to prophecy of a different type from that of the *piagnoni* or the predictions of the Duke's political detractors – and further, because of its assumed divine origin, why it was understood to be more 'true'. We may begin by looking at the natal chart for Cosimo I published by Francesco Giuntini in 1581 (FIGURE 4.2).[5] Giuntini was among the most prominent astrologers of the sixteenth century in Italy, and his is the most correct surviving chart with respect to the baptismal record of the birth.[6] Focusing first on the angles rather than the planetary positions we may note that Cosimo had the ascendant (ASC) at Capricorn 24° 16'. The mid-heaven, or *medium coeli* (MC) was nearly at Scorpio 20°. The descendant (DSC) and lower heaven (IC) are at Cancer 24° 16' and Taurus 20° respectively.[7] Thus, the Capricorn in the *impresa* represents the ASC. But what of the mid-heaven?

Following ancient precedent, Renaissance astrologers gave sidereal *paranatellonta*

---

[5] Francesco Giuntini, *Speculum Astrologiae* (Lyons, 1581), I, p. 127. Giuntini first commented on the chart in his single volume *Speculum Astrologiae* published in 1573, a year before Cosimo's sudden and untimely death in 1574 of a brain aneurism.

[6] Archivio del'Opera di Santa Maria del Fiore, Reg. Maschi, 1512–1522, fol. 140v (June, 1519). The document indicates the birth occurred on 11 June (Julian) 1519 at 1 2/3 hours past sunset, or 9:13 PM. Cf. Rousseau, *Cosimo I and Astrology*, pp. 70–72, n.17.

[7] Because of the importance of the comparison made by interpreters of the chart such as Giuntini and before him, Giuliano Ristori, to that of the Emperor Charles V, we may note that he also had the ASC in Capricorn 10° 25', and the MC at Scorpio 16° 54'. Cf. Rousseau, *Cosimo I and Astrology*, pp. 11–17, and related notes.

significant weight in interpreting any natal map. The long sections devoted to instructions on how to compute sidereal parallels in Renaissance astrological texts such as Giuntini's is evidence of the importance given them by contemporary astrologers. Giuntini's commentary on Cosimo's chart told of the dramatic results of the sidereal parallels found in it. The astrologer points out that he was especially favoured by the constellation Pegasus rising with the ASC. This, he says, connoted special fame and military glory over one's enemies. In addition, the wonderfully benefic stars of the Northern Crown had culminated over the mid-heaven in Scorpio. It was this that most accounted for his unexpected rise to power, his extraordinary success against his enemies, and his eventual coronation as the first Grand Duke of Tuscany.[8] It is evident then that the Northern Crown or Crown of Ariadne above the Sign of Capricorn in Cosimo's *impresa* specifically alluded to this constellation culminating above the mid-heaven degree.

This, however, was not all. Giuntini goes on to add that the degree of the mid-heaven in this chart was significant for another reason. At 20° Scorpio it was located in the degree of the 'third conjunction' – that is the conjunction of Jupiter and Saturn on 25 November 1484. This was the third great conjunction of Jupiter and Saturn in the watery triplicity since the *trigonalis* in 1365. The ninth-century Arab astrologer Abu Ma'sar had predicted that its mundane effect would include the end of the Christian religion. Fourteenth-century commentators such as Pierre d'Ailly and Paul of Middelburg tried to demonstrate how this was impossible since the origin of the Christian religion was divine. Nevertheless, the anxiety about the conjunction fed into millenarian fears about the year 1500. The attempt of Pico della Mirandola to critique the theory of the conjunctions, which was of medieval and not ancient origin, was in some part directed at this problem. Supporters and detractors of Martin Luther both put his birthdate at the time of the application of the conjunction, while others put it a year earlier. Luca Gaurico's chart (1552) shows his birth at 22 October 1484 with the conjunction in the ninth house – the house of religion – along with the Sun, Venus and Mercury. The astrologer adds a commentary referencing how Luther's life did indeed transform the Christian religion.[9] After the convulsions of the 1520s, the prophecies generated by the conjunction had made it so well known that Giuntini could just refer to it as the 'third conjunction' and assume his readers would understand.

––––––––––––––––––

[8] Giuntini, *Speculum Astrologiae*, 1573 and 1581 on Cosimo I's nativity. Cf. Rousseau, *Cosimo I*, Chapter I and related explanatory notes.

[9] Luca Gaurico, *Tractatus astrologicus in quo agitur de praeteritis multorem hominum accidentibus per proprias eorum genituras ad unguem examinatis…* (Venice: Bartholomaeus Caesanus, 1552) p. 69v.

Returning to the interpretation of Cosimo's chart, it was in the 1573 edition of his *Speculum Astrologiae* that Giuntini elaborated why having the third conjunction with the mid-heaven degree was so important in both Cosimo and Charles V's nativities. The text reads:

> I observed the geniture of the Emperor Charles V many times while he was still only the Duke of Austria and had not yet ascended the throne of Spain or the Empire… Wherefore I turned to the planets and looked closely at the doryphory of the Moon. And according to the rule of Ptolemy, the same Charles had the 20th degree of Scorpio in culmination, which very same degree was the location of the conjunction of Saturn and Jupiter in the year 1484. Now, the Jews say that in whichever nativity the degree of these conjunctions shall be found in the horoscope or in the culmination, that native shall be exalted, and raised to lofty station, and shall be a king… But to this it must be added that the nativity had a doryphory of the Moon and the shining star of the Crown at the culmination of the sky, and therefore he had two constellations: the Winged Horse in the horoscope, and the Northern Crown at the mid-heaven.
>
> The Grand Duke of Etruria, the unconquered, has a similar configuration. For Cosimo was also only a private citizen, of noble stock. For he indeed had these same two constellations – that is he had the degree of that same conjunction in the culmination of the sky, and he also had the Northern Crown and Pegasus. In addition to this, Mars, the Lord of the mid-heaven, is with the Basilisk in 21 degrees of Leo in his condition, and the Lord of the Ascendant [Saturn] is in his own house. And more than this is to be observed… since this same Cosimo (as everyone sees) shall increase his *imperium*. He made his enemies flee, and was victorious over them, and even greater things are still expected of him.[10]

Nothing really could have been clearer. The placement of the Crown of Ariadne above the sign of Capricorn representing Cosimo's ASC would have indicated the sidereal parallel to the mid-heaven. And for anyone who knew the culminating degree of the constellation, which is a constant, the Crown actually indicated the precise degree of the mid-heaven in Scorpio, well known to have been the location of the conjunction of 1484. With or without the accompanying motto, *ANIMI CONSCIENTIA ET FIDUCIA FATI (with a steady spirit and faith in my destiny),* it was a perfect image. Because of the pictorial variations in the representation of both zodiac sign and constellation, it was wonderfully versatile as well for use in countless works of art.

This most fortunate influence of the fixed stars and the degree of the conjunction

---

[10] Giuntini, 1573, bk. V, pp. 238v–239r, my translation. The reference to the 'Jews' is primarily to the Sephardic Messahalah and Abraham ben Ezra.. See Rousseau, *Cosimo I*, pp. 21–28 and notes 43–47 for a complete explanation of the references in this passage.

at the mid-heaven had long before Giuntini's texts been predicted directly to the Duke by the Carmelite astrologer Giuliano Ristori in a lengthy and complex prognostic that was completed, signed and dated 28 June 1537; that is, during the most difficult period of Cosimo's entire career.[11] Ristori's prognostic, although it featured a rectified nativity, is a document of primary importance. Its very existence is a testament to the Duke's belief in astrology. It proves that, finding himself in the midst of what must have seemed extremely unfavorable circumstances, he sought whatever indications the divinatory science could reveal about his present and future. The astrologer's comments cover a wide range of aspects, including the influences of fixed stars at the angles, and the degree of the conjunction, but also planetary aspects and dignities. At every turn, Ristori underlines the prognosis of greatness achieved, not without some struggle, but with unimagined success in the end. In his summary of the astrological evidence that 'promise to your Excellency greatness and royal dignity' (*promettono all' Ecc.a Vostra grandeza et degnità regali*) he ends with reference to the Northern Crown: 'And finally, he writes, 'the Crown of Ariadne, over the royal house [the tenth] predicts royal greatness for you. This constellation was given the name of Crown by the ancients because of the character of its effects'.[12] Surely this would have been a point of great consolation and reason for renewed self-confidence at that moment. A little over a month later, on 1 August, Cosimo would defeat the Florentine rebels at Montemurlo, be confirmed Duke at the end of September, and issue his astrological *impresa* not long after that.

There can be little doubt that, emboldened by the predictions of his astrologers, and by the evidence of recent events, Cosimo would have wanted to make his nativity public through other artistic means as well. Among the earliest expressions of this was the astrological program devised for the vault of the loggia of the Medici Villa at Olmo a Castello. This was commissioned from the artist Jacopo Pontormo in the fall of 1537 despite the cost – and this latter was not inconsequential.

Simply put, the tumultuous political situation in the city had severely undermined the economy and Cosimo's personal finances as well. The anti-Medici harangues of the new crop of *piagnoni* prophets always managed to draw a willing crowd of hearers in a city in which prophetic thinking had traditionally been linked to political

---

[11] Nativity and prognostic for Cosimo de' Medici by Giuliano Ristori, Florence, Biblioteca Medicea-Laurenziana, MS Plut. 89 sup. 34, fols. 133r–186r. For further details, see Rousseau, *Cosimo I*, Chapter 1, note 28.

12 '*Et finalmente, la Corona d'Arianna, tutte nella casa regale, regale grandeza le indovina. Sortì la predetta costellatione appresso gli antichi il nome di Coronoa mediante gli effete de' suoi significanti*', Ristori, 1537, fol. 167r; fols. 164v–165r on having the conjunction '*in culmine*' as a prediction of greatness.

events. With the real threat of Spanish occupation should the situation be stirred out of the young Duke's control, these preachers posed a real danger. Many in the business community, frightened of imminent disaster, were leaving the city if they could, while others refused to pay taxes to a government that could not, they echoed, by the will of God, last.[13] This only aggravated Cosimo's already ruinous financial situation. His predecessor Alessandro's extravagances had shamelessly depleted the public treasury, and the Florentine Senate, wishing to restrain the new Duke, voted him a pitifully small salary of 12,000 scudi a year. This barely satisfied his needs, and he had little in family reserves to supplement it. To make matters worse, Charles V had sent a contingent of Imperial troops from Milan to Florence to help keep a lid on open resistance, but their commander, the redoubtable Marchese del Vasto, insisted that Cosimo pay them! When he couldn't, the soldiers went wreaking havoc in the countryside, and their raiding was extremely serious. Yet, against all of this, Cosimo and his mother, Maria Salviati, commissioned a major fresco commission.

Pontormo had worked for the Medici Pope, Leo X, and it is possible that he had already represented features of Cosimo's nativity in a frescoed lunette at the old Medici Villa of Poggio a Caiano some twenty years previously.[14] Now, Cosimo would ask the same Pontormo to present his nativity far more boldly in the now tragically lost painted configuration of the planetary and probably stellar features of his natal chart in the loggia of the Villa Castello where he and his mother had lived in considerable penury following the death of his soldier father, Giovanni delle Bande Nere, in 1526. The loggia was the most public part of a villa, and had noble associations in Florentine tradition. As far as motive, beyond the political, there are interesting possibilities of astrological magic in the creation of a painted vault representing those positive influences observable in the sky on the evening of the young Duke's birth. Had Pontormo not taken nearly five years to complete the paintings, their impact would certainly have been greater, although important features, if not the entire map were most likely circulated by more direct means as well. There was clearly reason for Cosimo's chart, or at least the important aspects of it, to have been circulated in order to bolster the anti-*piagnoni* campaign that gave rise to the astrological *impresa* – particularly as it was popularly believed that Augustus had done the same.

---

[13] Rousseau, *Cosimo I*, pp. 33–38 and related notes.

[14] On Lorenzo de' Medici and the solar orientation of the Villa at Poggio a Caiano see Rousseau, 'Botticelli's Munich *Lamentation* and the Death of Lorenzo de' Medici,' *Konsthistorisk Tidskrift* 59, no. 4 (1990): pp. 79–86. On the possible representation of positions from Cosimo's nativity by Pontormo in one of the lunettes in the Grand Salone, see Rousseau, *Cosimo I*, pp. 145–52 and Cox-Rearick, *Dynasty*, pp. 214–20.

All previous discussions of the decoration at Castello have tried to focus on the confused and partial description of the ruined vault by Vasari in his *Life of Pontormo* written in the 1550s. The paintings, executed in oil on dry plaster rather than in true fresco, deteriorated so rapidly and were in such a ruinous state by the mid-1550s, that they were all but invisible. Still, what Vasari wrote indicates that the programme was astrological and featured planets with Zodiac signs, although he says nothing about the programme relating to Cosimo's chart.

> In the center of the vaulting then, he painted a Saturn with the Sign of Capricorn, and a Hermaphrodite Mars in the Sign of the Lion and of the Virgin, and some little Angels who are flying through the air like those of Careggi.[15]

We are fortunate in that Pontormo, for all his many faults, kept the great bulk of his drawings and studies, and a very large corpus of them has survived. Among these a small group can be associated with the figures painted at Castello. These are studies for the figures representing the planets accompanied by the zodiac signs in which they were located in Cosimo's nativity. [16]

Uffizi 6510F contains two studies of a nude boy for one or more of the flying *putti* mentioned by Vasari as similar to those of Careggi (FIGURE 4.3).[17] On the lower right we can see the figure of Saturn as an old man with Capricorn entwined around him. As in all the studies, Saturn is nude, but in the paintings the planetary figures were probably costumed or given identifying attributes. Saturn and Capricorn must have been placed in the vault in such a way as to suggest both the ascendant position

---

[15] Giorgio Vasari, *The Lives of the Most Excellent Painters, Sculptors, and Architects*, trans. G. de Vere, ed. P. Jacks, (New York: Modern Library, 2006), p. 333.

[16] Many of Pontormo's studies for this commission were first identified with Castello and published by Janet Cox-Rearick (*The Drawings of Pontormo* [Cambridge, MA: Harvard University Press, 1964]) who was also the first to suggest that the vault may have contained a pictorial rendering of Cosimo's chart. Unfortunately, her later reading of the program (*Dynasty*, pp. 258–69) is still based on Vasari's text and is further confused by her attempt to suggest that the transit chart of the day Cosimo was elected by the Senate to become Duke was merged with the nativity. This is certainly incorrect. I have previously presented an alternate interpretation and selection of the surviving drawings in Rousseau, *Cosimo I*, pp. 301–19. Other attempts were made by David Wright ('The Medici Villa at Olmo a Castello: Its History and Iconography', PhD Diss., Princeton University, 1976) and L. Châtelet-Lange ('The Grotto of the Unicorn in the Garden of the Villa at Castello,' *Art Bulletin* 50, 1968, pp. 51–58) neither of whom saw connections to the Duke's natal chart.

[17] Vasari, *The Lives*, pp. 331–32. About the work in the loggia of the Villa at Careggi for Alessandro I which has also not survived Vasari noted, 'some little boys (*putti*) that were going in the oval space of the vaulting with various animals in their hands, and all foreshortened to be seen from below.' The possibility of the 'animals' being zodiac signs is very intriguing, yet no other evidence survives.

of the sign and the planet Saturn located just over the eastern horizon. A figure in another drawing now in the Fogg Art Museum (1932/144) has been convincingly identified as a development of the study of Saturn in the Uffizi sketch.[18]

As mentioned by both Ristori and Giuntini, the planet Mars was particularly well placed in Cosimo's chart in the sign of Leo parallel to its largest star, Regulus, the 'Lion's Heart'; a placement that may also have conjured the Florentine Lion mascot, the *Marzocco*. Uffizi 6760F is certainly a study for Mars in Leo (FIGURE 4.4). In this drawing, Mars relaxes while he leans against a lion lightly indicated under the god's right arm. This figure would have been found in the vault opposite and somewhat above Saturn in Capricorn.

The Sun in Gemini is represented in an exceptionally delicate drawing (Uffizi 17405F) representing Apollo as the Sun, holding a curved bow in his left hand (FIGURE 4.5). Almost prone, the figure of Sol-Apollo turns back to gaze upward at the configuration of planets that must have been shown setting after him as the Sun is the only planet in Cosimo's nativity located below the DSC horizon. The planets Venus and Mercury and the Part of Fortune were conjunct in the sign of Cancer in the seventh house of the chart. It appears that Pontormo struggled to accommodate all these figures with the rather unwieldy lobster-like creature that was the common representation of the sign in the fifteenth and sixteenth centuries. FIGURE 4.6, printed in 1535, is a good example of illustrations of the sign that almost certainly derive from a series of simpler woodcuts used in all astrological texts published by Ratdolt since the late fifteenth century.[19] We can see a similar representation of Cancer in the famous star map drawn by Albrecht Dürer printed in 1515 (FIGURE 4.7).

We can compare this rendering of the sign to Uffizi 700S (FIGURE 4.8) which I here propose as a set of first ideas to accommodate the knotty quartet in Cancer. A page of red-chalk sketches, like the others in the group, this drawing is carefully divided down the middle by a ruled line. On the larger right hand side we can see a barely decipherable female nude that seems to extend her hand toward the first study of the zodiacal creature. This is followed by two further attempts at drawing the crustacean, and a flying putto. Ignoring the mask which does not seem related to the other drawings on the page, we may direct our attention to the bottom right hand corner. Here we can see a pair of reclining figures, one with its back to the viewer that appears male, and another facing him that appears female. Above them is the very faint beginning of another female figure that, in this context, would have represented

---

[18] Cox-Rearick, *Drawings of Pontormo*, I, cat. 336.

[19] Cancer, Leo, Virgo from Hyginus, *C. Iulii Augusti Liberti Fabularum Liber* (Basel: Ioan. Hervagium, 1535), p. 96.

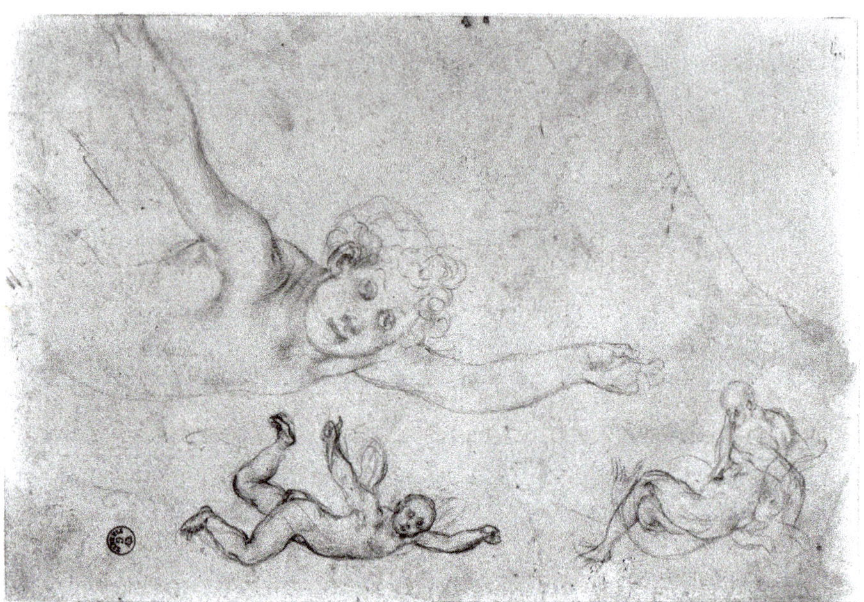

FIGURE 4.3: Pontormo, Uffizi 6510F, Study for Castello vault, flying *putti* and Saturn in Capricorn

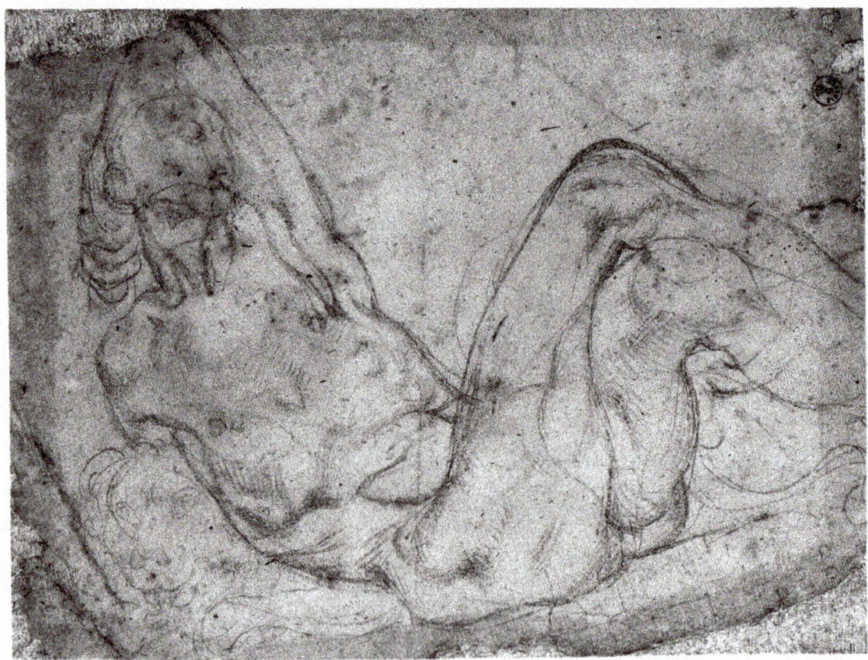

FIGURE 4.4: Pontormo, Uffizi 6760F, Study for Castello vault, Mars in Leo

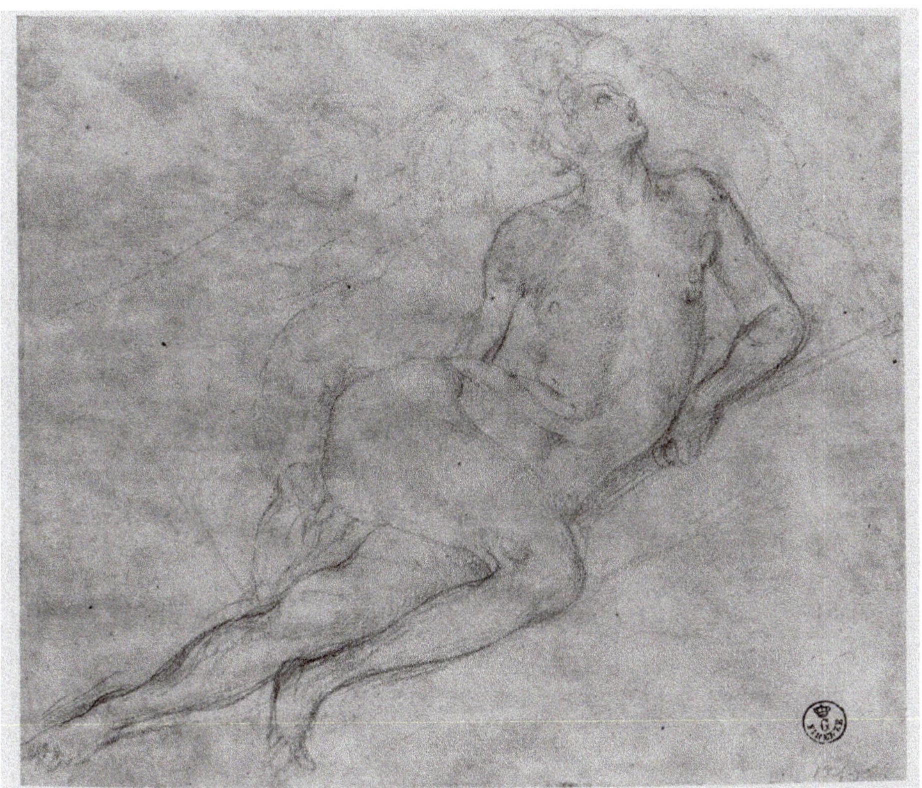

FIGURE 4.5: Pontormo, Uffizi 17405F, Study for Castello vault, the Sun in Gemini

Fortuna as the Part of Fortune in the chart.

Uffizi 6630F, another striking red-chalk drawing by Pontormo (FIGURE 4.9), may represent a clue to the final artistic solution to the problem of so many figures in one place. Vasari mentions having seen a 'hermaphrodite' in the vault, and this drawing certainly seems to represent such a figure. It may well have been pointed out to the artist that Mercury is a planet of common gender; masculine in nature with masculine planets, and feminine when found with the Moon or Venus, especially when he is vespertine as in the planet's position in Cosimo's chart. If informed of this option, it would have made sense for Pontormo to decide to represent the conjunction in Cancer with a Hermaphrodite, the study for which can be seen in Uffizi 6630F. Although subtlety was not an important priority at Castello, I believe that the compositional difficulties presented by the grouping of Fortuna, Venus, Mercury and the lobster-like Cancer under Mars in Leo were indeed solved by placing the Part of Fortune as a definite female figure between Mars and the Hermaphrodite

Cancer.

Cancer nullam habet insignem stellam, sed omnes sunt uel quartæ uel quintæ magnitudinis.

Vnc medium diuidit circulus æstiuus ad leonis exortus spectantem paulu lum supra caput hydræ collocatum. Occcidentem & exorientem posterio re corporis parte. Hic autem habet in ipsa testa stellas du as, quæ asini uocantur, de quibus ante diximus, in pedibus dextris singulas obscuras, in sinistro pede primo duas, in secundo duas obscuras, in tertio unam, in quartoprimo unâ obscuram. in ore unam, in ea quæ chela dexterior di citur tres similes, non grandes, in sinistra similes duas. Et ita est omnino stellarum numerus decem & octo.

Leó.

Leo habet duas stellas primæ magnitudinis, quarû una uocatur Calbalezet, id est, côr leonis, alia Deneb ale zet, hoc est, cauda leonis. Habet tres secundæ magnitu dinis, duas in ceruice & unam in dorso, reliquæ sunt tertiæ, quartæ, quintæ, & sextæ magnitudinis.

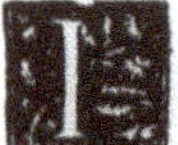Eo spectas ad occasum supra corpus hydræ à capite quà câ cet instat usq ad mediâ parte eius côstitutus, medius æstiuo circulo diuiditur, ut sub ipso orbe priores pedes habeat collocatos, occidens à capi te & exoriens, hic habet in capite stellas tres, in ceruicibus duas, in pectore unâ, in interscapilio tres, in media cauda unâ, in extrema alterâ magnâ, sub pectore duas, in pede priore unam claram, in uentre claram unam, & infra alteram magnam unam, in lumbis unam, in posteriore genu unam, in pede posteriore claram unâ. Et ita est omnino numerus stellarum decem & nouem.

Virgo.

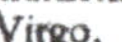Virgo unâ habet stellam primæ magnitudinis, uocaturq ea Azimech, id est, spica uirginis, habet deinde sex tertiæ magnitudinis, tres in ala sinistra, unam in latere dextro, unam in manu dextra, et unam sub ala dextra, reliquæ sunt quartæ, quintæ & sextæ magnitudinis.

Irgo infra pedes Bootæ collocata, capite posteriore parte leonis dextra manu cir culum æstiualem tangit, ac inferiorem partem corporis supra cotuû & hydræ caudam habere perspicitur. Occidens ca pite priusquam reliquis membris, huius in capite est stella una obscura, in utrisq humeris singulæ, in u trisq pennis binæ, quarû una stella quæ est in dextra pena ad humerû defixa protrygeter uocatur. Præterea habet in utrisq manibus singulas stellas, quarû una quæ ex in dextra manu, maior & clarior conspicitur, in ueste autem habet passim stellas septe, in utrisq pe dibus singulas. Omnino stellarum numerus xvi.

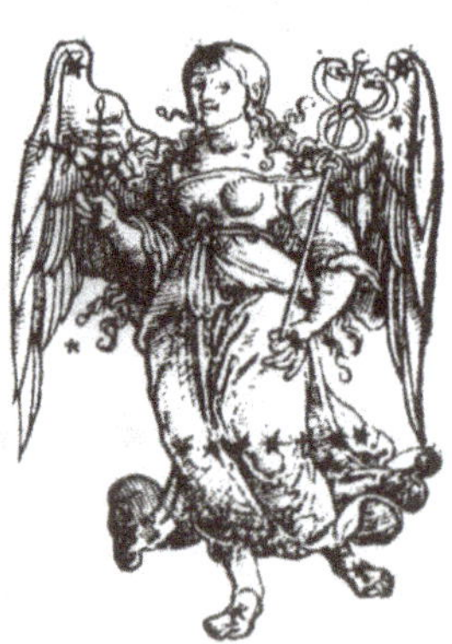

FIGURE 4.6: The signs of Cancer, Leo and Virgo, from C. *Ivlii Hygini Augusti Liberti Fabularum Liber…*Basel, 1535.

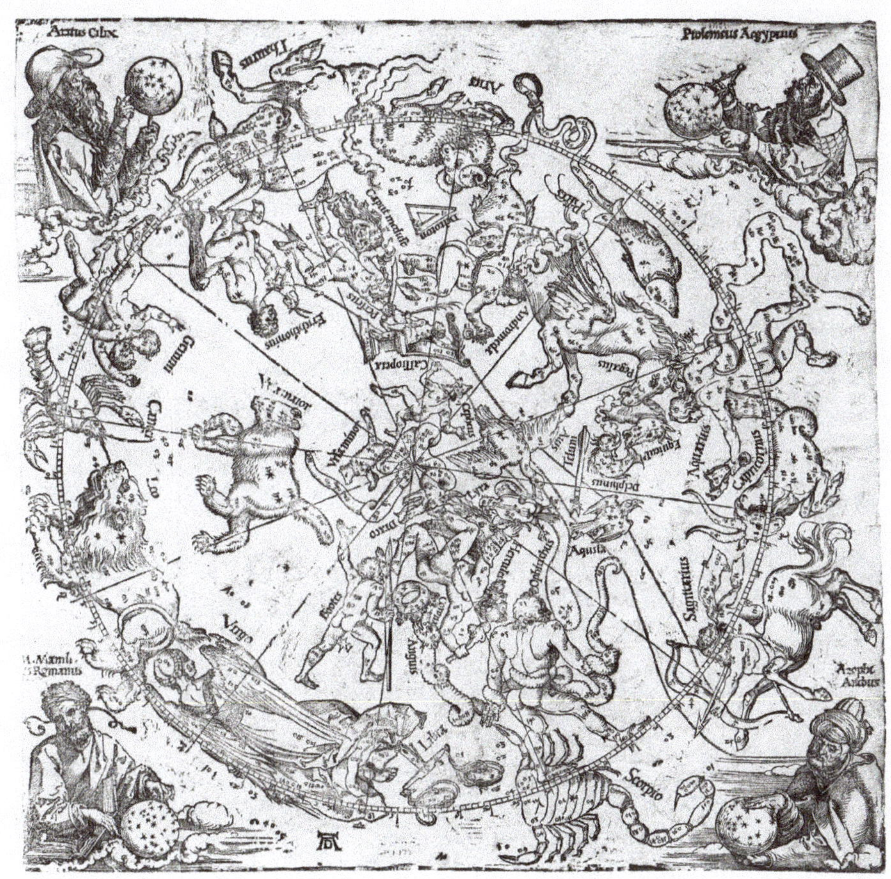

FIGURE 4.7: Albrecht Dürer, with Johannes Stabius and Conrad Heinfogel, Star map of the northern hemisphere, 1515

with Cancer. This would have maintained the relative aspects of their positions in Cosimo's nativity. This arrangement would also help us to make some sense of Vasari's description of the ruined paintings in which the Crab may no longer have been visible. Counting the wrong way through the Zodiac, Vasari guessed that the figure that was actually Fortuna, and which may have been winged, and which was located in some proximity to Mars, was the Sign of the Virgo. This Sign, which is often also seen represented with wings, lies on the other side of Leo. Hence Vasari's confusion about the placement of Mars in the vault.

These complete the positions that we know were represented in the vault, but it would make sense to imagine the rest of the planetary positions, as well as the all-

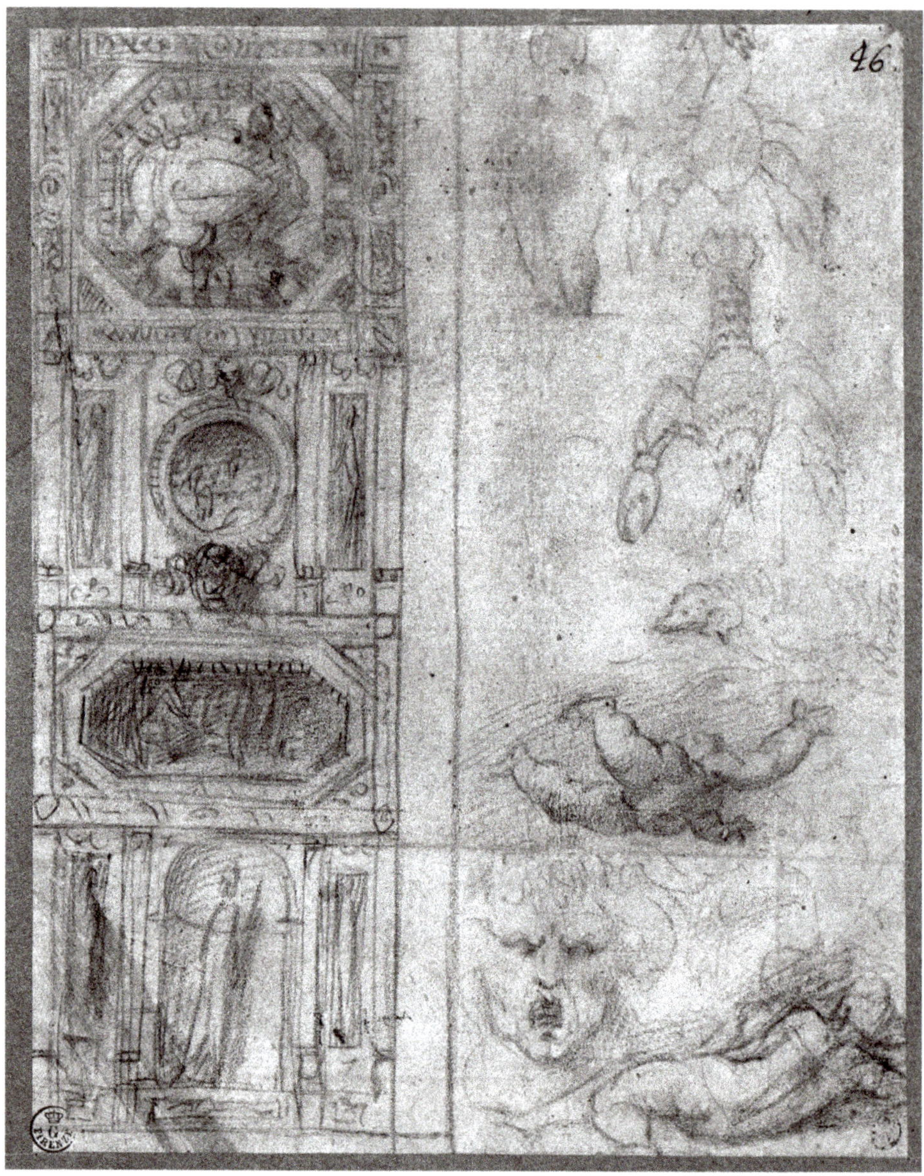

FIGURE 4.8: Pontormo (attributed), Uffizi 700S, Sheet of studies for the conjunction of Mercury and Venus in Cancer, with the Part of Fortune for the Castello vault.

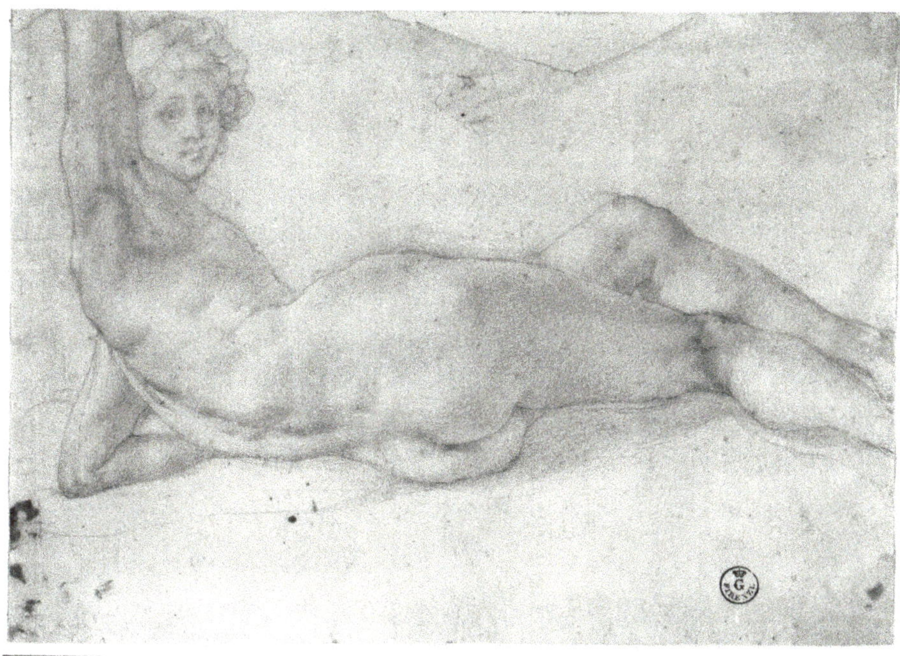

FIGURE 4.9: Pontormo, Uffizi 6630F, Study for Castello vault, Hermaphrodite representing the conjunction of Venus and Mercury.

important constellation of the Crown of Ariadne, either as a crown, a wreath, or a ring of stars.[20] It may well have been shown as the constellation, particularly because it was already a part of the Duke's *impresa* in that form.

The choice of an entrance *loggia* for an important painting commission of this kind had a special public significance in the sixteenth century. *Loggie* had been traditionally used for entertaining important guests on special occasions and for this reason were typically the location of decoration that celebrated the family or its current head. The aristocratic suggestions of a loggia setting are suggested by the expression *famiglie di loggia* (loggia families) that was applied to the noble families of Florence who, before the thirteenth century, were the only ones who were permitted to have a loggia that faced out onto a public thoroughfare.[21] The young Duke and his mother would have been keenly aware of the social meaning of a decorated loggia. Despite the extravagant cost of such a venture, the refurbishing of both the house and the

---

[20] On the pictorial tradition of representation of this constellation, see Rousseau, *Cosimo I*, Appendix A.

[21] Wright, *Medici Villa*, pp. 261ff, n. 189.

garden at Castello with an elaborate program that glorified Cosimo I was intended as a showpiece of his importance that was especially appropriate since the country villa had traditionally been the locus for public display of this kind.[22] The luxury implied by the enlargement and redecoration of his villa projected an image of wealth and power, neither of which, as discussed above, Cosimo had in abundance in his early years. Despite these difficulties, the plan was initiated, and while Pontormo was working there, the villa was quite evidently in continuous use. By 1543, the year in which the desperate Maria Salviati ordered that the paintings in the loggia finally be revealed to the public, Cosimo was maintaining a household at Castello that numbered as many as 348 people.[23] Maria and Cosimo's frustration with Pontormo's slowness in completing the work is understandable in light of the propagandistic purpose that lay behind the astrological decoration in the first place. By 1543 the pressing need for it was certainly less great than it had been five years previously, but its unveiling did coincide with a series of gestures on Cosimo's part that demonstrated the fulfillment of the augury of his nativity, such as taking possession of the Florentine fortresses. It may, therefore, be presumed, that the intent and desired effect of the decoration was not lost on those who viewed it. One can only speculate on how the Duke may have felt about the rapidly deteriorating condition of the paintings so soon after their long-awaited completion. In any case, by the early 1550s Cosimo seems to have lost interest in the Castello project, as the sculptural plans for the garden were left only partially carried out. By this time the Duke was concentrating his attention on the symbolic decoration of the Palazzo Vecchio into which he had moved in 1540 in a decisive political move to take over the seat of Republican Florence. The Palazzo Vecchio is replete with references to his astrological destiny, most evidently in the ubiquitous representation of the Capricorn and stars emblem. Used in almost a talismanic fashion on the doors and walls, it was also embedded in the final programme of the largest room in the Palazzo (FIGURE 4.10) which, as Castello was to have done, celebrated Cosimo's glorious predestined reign, and the golden age he had brought to Florence.[24]

Evidently then, the painted nativity in the loggia was a kind of artistic publication of the propitious positions of Cosimo's chart and, as propaganda, would have been understood by the audience he intended. However, quite apart from these facts, we might speculate on other motives for having it made. Mary Quinlan-McGrath has recently discussed evidence, particularly in the writings of Marsilio Ficino, whose

---

[22] Wright, *Medici Villa*, pp. 33–39; 40ff.

[23] Wright, *Medici Villa*, pp. 62–63.

[24] On this see Rousseau, *The Pageant…* pp. 417–18, 423–30.

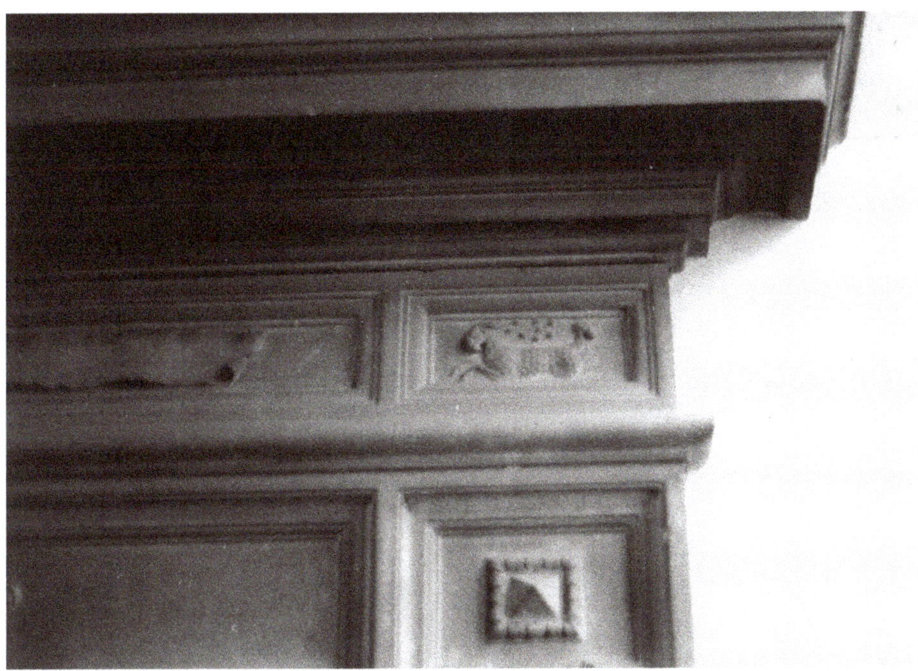

FIGURE 4.10: Capricorn and Crown, Door decoration, Palazzo Vecchio, c. 1550s.

patrons had been the Medici family, that such a representation would have been believed to potentially pull the influences of those stars and planets in their well-placed
positions down to the native, in this case, the Duke himself. It was better than the
actual sky, which would never be in exactly those positions again. It was a permanent
sky that radiated the positive influences that someone like Cosimo would have felt he
sorely needed at the time he commissioned the vault. Ficino recommended staying
under images of this kind, even sleeping under them.[25] The bases for these connections are found most extensively in the third book of Ficino's *De vita*.[26] Nevertheless,
the idea of correspondences between the heavens and things in the environment is
the origin of most of the ideas in this work published in 1489. Interestingly, in the
passage cited above, Ficino refers to his 'friend' called 'Lorenzo' – almost certainly
Lorenzo de' Medici, il Magnifico, Ficino's patron at the time – who had just ordered

---

[25] Mary Quinlan-McGrath, *Influences: Art, Optics and Astrology in the Italian Renaissance* (Chicago:
University of Chicago Press, 2013), especially chapters 7 and 8.

[26] Marsilio Ficino's *De vita coelitus comparanda*, Book III: 'On Obtaining Life from the Heavens',
chapter 20, pp. 44–58 (Marsilio Ficino, *Three Books on Life*, Carol V. Kaske and John R. Clark, eds.
and trans., The Renaissance Society of America, Binghamton, NY, 1989).

a painting of this kind. Presumably once in the Palazzo Medici, it is no longer extant.

The doctrine of correspondences which suggested that images could hold this kind of power may well have been a primary motive for other representations of this kind in the Renaissance. A good example is Peruzzi's ceiling that represents the nativity of Agostino Chigi in the *Sala di Galatea* in 1511–12.[27] Or, the Sala dei Pontefici vault painted for Pope Leo X in 1520–21.[28] And, it might be imagined to have been important in Cosimo's decision to do the same in 1537.

As mentioned earlier, the astrological *impresa* was seen in all manner of artistic representations, with increasing occurrence after Cosimo had actually assumed political power. His belief in the power of the stars hardly waned through his career, and as I have shown, there is evidence that he based his strategy in war on astrological prognostics which he continued to receive throughout his lifetime.[29] The messianic quality of his propaganda did not change. For example, an allusion to his astrological fortune was suggested in a census book where the Dove of the Holy Spirit is seen radiating on the Capricorn and stars with a text reminding the viewer that Cosimo was sent by God to care for Florence and to '…bring to a perfect life this Excellent State'.[30]

There were, however, other interesting variants, always alluding to *true* prophecy as opposed to the *false* prophecies of the *piagnoni* and others against him. In one of these, alluding to the theme of true prophecy, we see the adaptation of the figure of Apollo *arcitenenes*, the same Apollo who was represented at Castello as the Sun. Using this image, Cosimo developed a new impresa as represented in a medal issued

---

27 Baldassare Peruzzi with Sebastiano del Piombo, ceiling fresco in the so-called *Sala di Galatea* (after Raphael's fresco of that subject on the lower wall) in the Villa Farnesina in Rome. The first author to consider the fresco as a representation of Chigi's horoscope was Fritz Saxl (*La Fede Astrologica di Agostino Chigi: Interpretazione dei dipinti di Baldassare Peruzzi nella Sala di Galatea della Farnesina*, Rome, 1934). Since then there have been a number of important studies of this complex, notably Mary Quinlan-McGrath, 'The Astrological Vault of the Villa Farnesina: Agostino's Rising Sign,' *Journal of the Warburg and Courtauld Institutes* 47 (1984): pp. 91–105; Ingrid Rowland, 'The Birthdate of Agostino Chigi: Documentary Proof,' *Journal of the Warburg and Courtauld Institutes* 47 (1984): pp. 192–93; Kristen Lippincott, 'Two Astrological Ceilings Reconsidered: The Sala di Galatea in the Villa Farnesina and the Sala del Mappamondo at Caprarola,' *Journal of the Warburg and Courtauld Institutes* 53 (1990): pp. 185–207 and Quinlan-McGrath, *Influences*, pp. 172–80.

28 On the Sala dei Pontefici in the Vatican Palace see Rousseau, *Cosimo I*, pp. 153–77; Cox-Rearick, *Dynasty*, pp. 188–98, and Quinlan-McGrath, *Influences*, pp. 180–89.

29 For a striking example, see Rousseau, 'An Astrological Prognostic,' citing the critical prognostic received by Cosimo I from Francesco Formiconi written on 27 March 1554, the year of his victory over the Florentine exiles led by Piero Strozzi and his conquest of Siena.

30 Census book made for Cosimo I by Antonio di Filippo Danti Gianetti completed 25 February 1552 (Florence, Archivio di Stato di Firenze, Medicea Misc. 223, c. 7v. The explanatory text is on cc. 3v–4r.

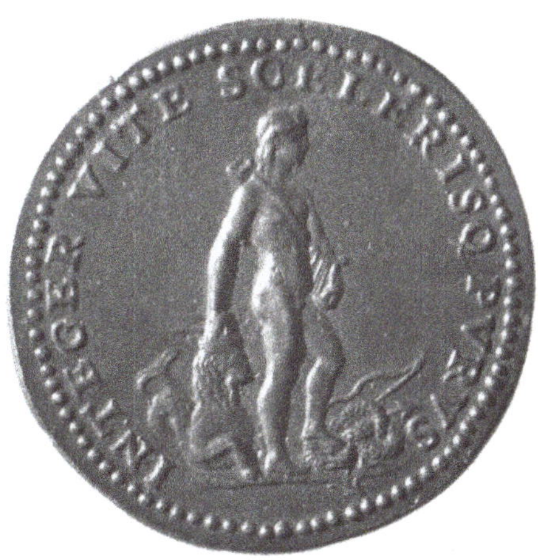

Figure 4.11: Domenico Poggini, Medal for Cosimo I de' Medici, *Apollo Pythonicida*/Capricorn/Northern Crown/INTEGER VITE SCLERESIQ[UE] PVRVS, 1554–55, Museo Nazionale del Bargello, Florence

in 1554–55 (Figure 4.11) by Domenico Poggini.[31] *Apollo pythonicida*, Apollo, killer of the dreadful Python with his arrows, gave him prophetic powers. This issue of the medal coincided with the Duke's spectacular victory over Siena in 1554, with strategy based on his astrologer's advice. In Cosimo's medal, the Apollo of 'true prophecy' crowns Capricorn with the Northern Crown. A few years later, the same artist enlarged this idea in a sculptural representation made for the Boboli Gardens (Figure 4.12) where we see Apollo crowning Capricorn with the Northern Crown.[32]

Yet this insistence on the evidence of the fulfillment of the astrological prophecy never waned in Cosimo I's imagery. Even after he had been crowned Grand Duke, a new position created for him by the papacy instead of the kingship that should have been rightly his, a medal of 1570 shows that he still believed in the influence of this same constellation in this last great honour that was bestowed upon him.[33] His profile with the Grand Ducal crown is backed by the constellation of the Crown of Ariadne behind his head (Figure 4.13).

Created originally in opposition to the constant 'wailing' of the *piagnoni* (i.e., *the wailers*) about God's will that Cosimo would fail, through works of art the Duke

---

[31] Medal for Cosimo I, Apollo Pythonicida/Capricorn/Northern Crown/ INTEGER VITE SCLERIS[QUE] PVRVS, 1554–55, Bargello, Florence.

[32] Domencio Poggini, *Apollo Pythonicida/Medicus Crowning Capricorn with the Northern Crown*, marble, formerly with bronze additions, 1559, Boboli Gardens, Florence.

[33] Francesco da San Gallo, Medal for Cosimo I, 1570, Museo Nazionale del Bargello, Florence.

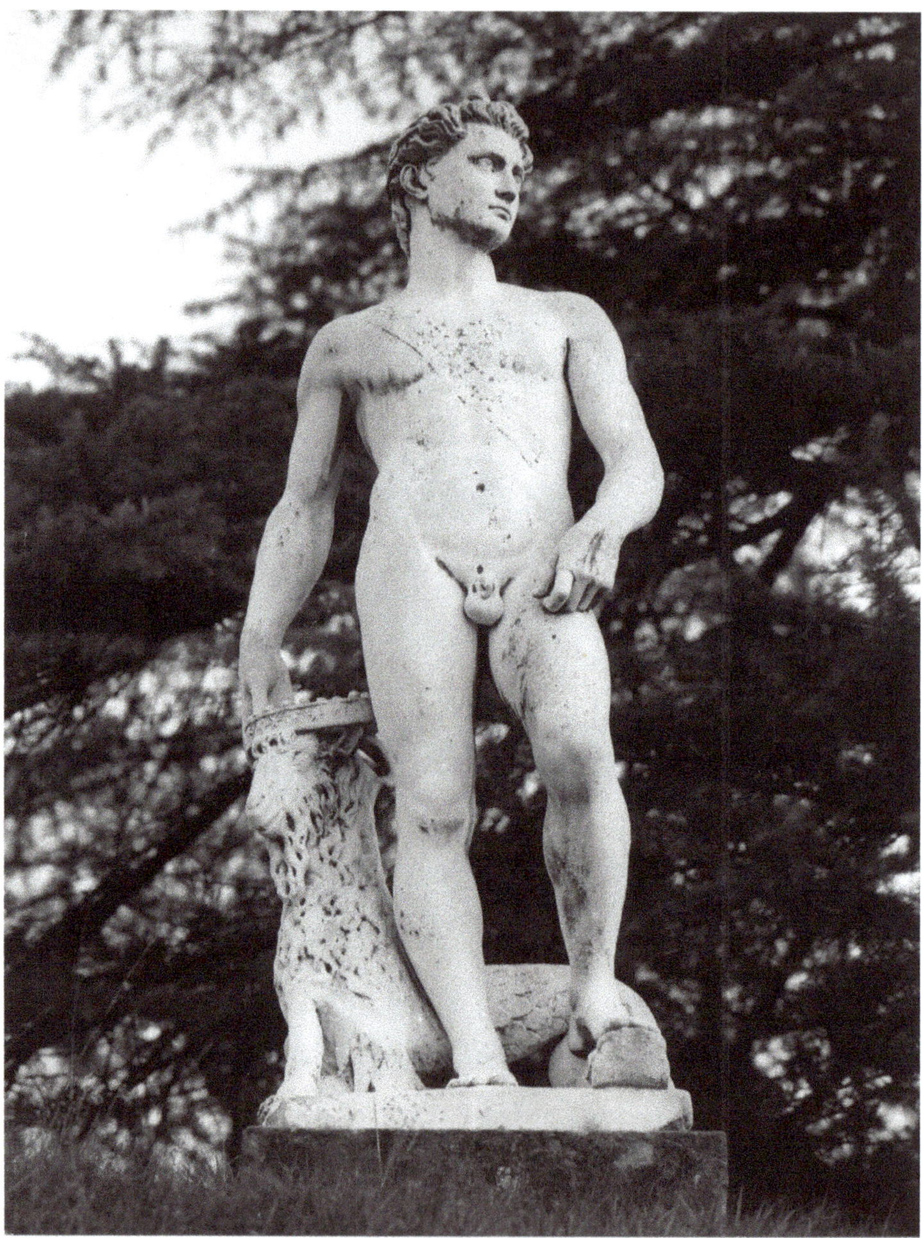

FIGURE 4.12: Domenico Poggini, *Apollo Pythonicida/Medicus Crowning Capricorn with the Northern Crown*, marble, formerly with bronze additions, 1559, Boboli Gardens, Florence.

continued to promote evidence of what he believed to be a truer prophecy. That prophecy came from the stars.

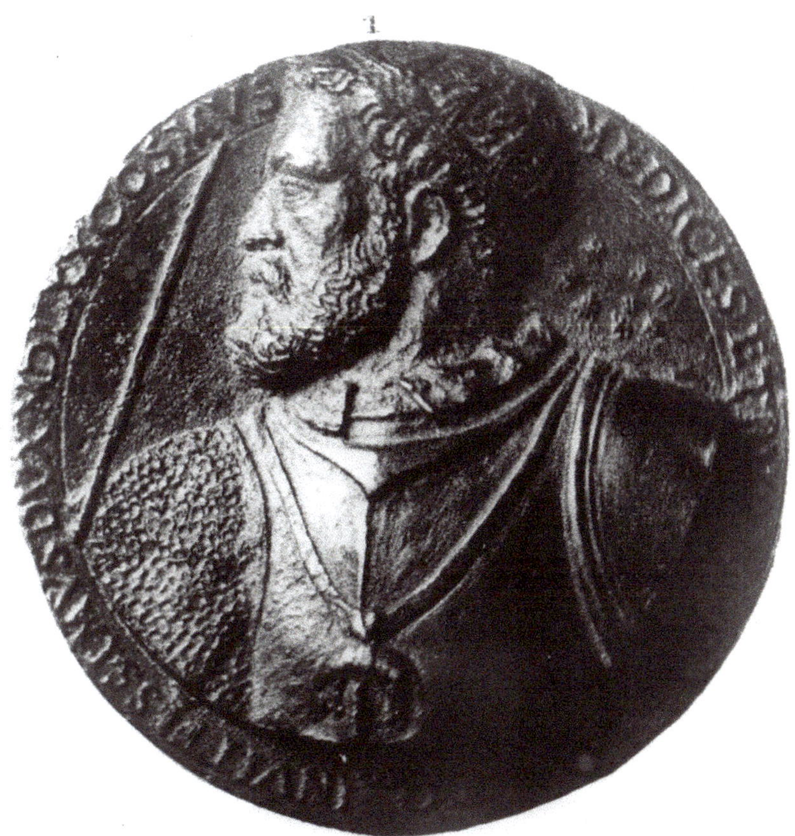

Figure 4.13: Francesco da San Gallo, Medal for Cosimo I de' Medici as Grand Duke of Tuscany, with the Northern Crown, 1570, Museo Nazionale del Bargello, Florence.

Astrological Symbolism in Wolfram von Eschenbach's Parzivâl

John Meeks

ABSTRACT: Unlike the French Grail romance of Chrétien de Troyes and his successors, Wolfram's *Parzivâl* is enriched by vivid astrological motifs and symbols. Some of these are explicit, while others seem to be hidden. Wolfram's enigmatic terminology, and his account of how the name of the Grail was first read in the stars, have given rise to much speculation and controversy. This chapter offers a possible interpretation of the astrological background of the narrative, in particular of the role played by Saturn. I argue that in the course of Parzivâl's inner development, he is able to overcome that planet's malevolent influences by acquiring its virtues. Only then can he successfully ask the question of compassion which heals the suffering Grail King.

Wolfram von Eschenbach, the celebrated German *Minnesänger* and epic poet of the early twelfth century, has become something of a legend himself, and if an aura of mystery had already come to surround him soon after his death in 1220, there are enigmas enough in his works to account for it.[1] He insists that he could neither read nor write, although perhaps no other vernacular poet of his time demonstrates such wide-ranging interest in different cultures, languages and religions. No other thirteenth-century poet was so cosmopolitan in the subjects he treated, or so tolerant in his attitude to non-Christian religions. Even more puzzling is his relationship to astrology and astronomy. His best-known epic, *Parzivâl,* usually regarded as an elaboration of the unfinished tale of Chrétien de Troyes, adds a completely new dimension by including a macrocosmic background to the events surrounding the Grail and the suffering of the wounded Grail King. Wolfram stresses the importance of the tale's celestial background, and his complaint that 'Master Christian of Troys has done the story an injustice' may well be aimed at this omission.[2]

Was Wolfram himself knowledgeable in the science of the stars? Or did he transmit the knowledge he received from some other source? This question was taken up by one of the most widely-read works in thirteenth century Germany: *The War of the Minstrels at the Wartburg.*[3] This anonymous collection of poems and riddles, which

---

[1] Little is known historically of Wolfram's life. On the legends surrounding him see Dieter Kühn's commentary in Wolfram von Eschenbach, *Parzivâl,* original text and translation by Dieter Kühn (Frankfurt a.M.: Deutscher Klassiker Verlag, 1994), p. 413.

[2] '...von Troys meister Cristjân / disem maere hât unreht getân...', *Parzivâl,* p. 827, 1–2.

[3] *Der Wartburgkrieg,* edited and translated by Karl Simrock, (Stuttgart & Augsburg, 1858). The English translations from the *Wartburgkrieg* are my own.

appeared some twenty or thirty years after Wolfram's death, describes a contest among the most famous minstrels, which is said to have taken place in 1207, at which time Wolfram may still have been at work on *Parzivâl*. In the first part, 'In Praise of Princes', the poets vie with each other in songs dedicated to the glory of the princes they serve. Heinrich von Ofterdingen is bettered by his opponent Walther von der Vogelweide, who employs a cunning rhetorical device to win the laurels. As in the fabled contest between Apollo and Marsyas, the loser must pay with his life, and so we need not wonder that Heinrich is not pleased with the outcome. He appeals to the jury to allow him to bring his master, the Hungarian sorcerer Klingsor, to continue the contest into a second round on his behalf. His plea is granted, and when Klingsor arrives, he engages Wolfram von Eschenbach in a 'Contest of Riddles'. Wolfram now finds himself face to face with the very necromancer he created in *Parzival*, but never allowed to appear personally in the story, letting him cast his spells from a safe distance. Wolfram, however, is equal to the test, and baffles the magician by solving the most arcane riddles he is able to pose. When at last Wolfram shows his knowledge of the fourfold symbolism of the Evangelists as ox, lion, eagle and man, Klingsor suspects that he must be aware that these are the four cardinal constellations of the ancient zodiac. Frustrated, he cries out, 'Whoever thinks you are a layman is not in his right senses. Astronomy is in league with you! If you won't tell me yourself, the devil Nasion will find you out this very night!'[4] He threatens to summon this devil from Toledo, well-known as a meeting-place between Arab and Christian culture, and as an important centre of knowledge in astronomy and astrology. Nasion, coming to Wolfram in his sleep, asks a few questions to test his knowledge: 'Now tell me, if you are a master, how it is that the firmament with its lofty power strives against the planets?', and further, 'When Saturn stands in the east, what is the meaning of this wonder?'[5] Both questions are clearly related to themes Wolfram treats in Parzival, and are certainly well chosen to discover whether or not its author was speaking out of his own knowledge. The first question is astronomical, the second requires an understanding of astrology. When, near the end of *Parzivâl*, the sorceress Cundrie lists the names of the planets in Arabic, she says that they strive against the movement of the firmament, meaning that its diurnal movement from east to west is slowed down by the planets' counter-movement from west to east: 'die sint des firmaments zoum, die enthalden sine snelheit', 'They are the bridle of the firmament, they strive against its speed'.[6] Nasion reverses the imagery, letting the firmament strive against

---

[4] *Der Wartburgkrieg*, p. 107.

[5] *Der Wartburgkrieg*, p. 110.

[6] *Parzivâl*, p. 782, 15–16. All English translations of *Parzivâl* are my own.

the planets. The question concerning Saturn would have reminded those familiar with *Parzivâl* that when this planet reached its 'goal', the pain of Anfortas' wound was greatly intensified. Despite having explicitly expounded on this counter-movement in *Parzivâl*, and despite his frequent references to Saturn there, the Wolfram of the *War of the Minstrels* readily confesses his ignorance, saying, 'As far as the virtues of the planets, the courses of the stars, and the music of the spheres are concerned: all I know is that the Almighty laid out their courses, both for night and day'.[7] Nasion, upon hearing this, is furious to think that such an ignoramus should have caused him to travel so far in vain. His contempt for Wolfram knows no bounds, and the poet has no choice but to drive him forth with the sign of the cross. This earns him some respect from Nasion, who reports back to Klingsor that Wolfram is 'so clever, that you will never stand a chance against him'.[8] Such was the answer offered in the earliest version of the *War of the Minstrels* to the question of Wolfram's knowledge of the stars.

I propose to take a fresh look at this question, and to do so, it will be necessary to recount some of the most important stages of Parzival's quest. Inevitably, many significant and poetical episodes will have to be omitted in favour of those which seem most relevant to the question at hand.

Parzival's childhood is spent in the solitude of a remote forest. Here his mother Herzeloyde hopes to keep him safe from all knowledge of knighthood, for she fears that he will meet with the same fate as his father, who lost his life in battle. But one day while out hunting, the youth encounters a group of knights whose armour is so resplendent in the sunlight, that he believes each one at first to be a god. Returning to his mother, he announces his intention of becoming a knight of King Arthur, having learned of his Round Table from the knights in the forest. Herzeloyde sends him forth in fool's clothes, in the hope that he might become a laughing stock, and soon return home discouraged. Parzival rushes forth without so much as turning back for a last greeting, and his mother falls lifeless to the ground. Oblivious that she has died of grief for his sake, yet blindly obedient to every piece of advice she had given him, he commits many follies, and inadvertently causes suffering through his foolish conduct.

When he first approaches the place where Arthur and his court are encamped, he encounters a knight clad all in red. So fascinated is Parzival with the beauty of his armour, and so greatly does he covet it, that he can think of nothing else but acquiring it. After delivering a message from the knight to King Arthur, Parzival has the

---

[7] *Der Wartburgkrieg*, p. 111.
[8] *Der Wartburgkrieg*, p. 114.

audacity to ask for the knight's armour. The king hesitates, then tells him he will have to win it for himself, and with this assurance, Parzival rides back to where the knight is waiting. He repeats his demand, and when the knight, to reprove him for his insolence, strikes him to the ground with the butt end of his lance, Parzival responds by sending his *javelot*, the little spear he had brought with him from his forest home, flying through the visor in the knight's helmet, killing him at once.

Up until this moment, Parzival has been guided in his actions by a kind of child-like foolishness, and whatever havoc he may wreak, he does so innocently. But when he strikes down the Red Knight, he does so in a sudden rage. He shows himself to be not only fearless, but impulsive and hot-tempered. The news of the knight's death causes great consternation in Arthur's court. The sorrow over his death seems nothing less than a tragic foreshadowing of the sorrow Parzival's visit to the Grail Castle will soon cause. He rides away as the 'Red Knight', but beneath the armour he still wears the motley garb of the fool, given him by his mother. His courage, together with his impulsiveness and his sudden outburst of temper, allow him to win the red armour, which he wears all through the further course of his adventures. Although it is never explicitly stated, these qualities correspond closely with those associated with the planet Mars. Astrologers working with the medical tradition of the four humours connected the planet's red colour to 'red bile', which was thought to be responsible for the choleric temperament.[9]

Taking his leave of Arthur's court, Parzival soon arrives at the home of the grey-haired Gurnemanz, who imparts to him much-needed instruction in the arts of knighthood and courtly decorum. He is acquainted with a list of virtues, each one of which is described in terms of a polarity between opposites. Thus, he learns to be bold, yet merciful, and generous without squandering his wealth. Man and woman, he is told, are as one, and blossom from a single seed. Above all, Gurnemanz entreats Parzival not to ask too many questions, but to learn to listen also to what others may wish to learn of him. Besides, much can be discovered through patience and close observation. In this way Parzival is able to cast off his childish ways and his unthinking obedience to his mother's advice. Yet the advice of Gurnemanz will only avail him if he is able to apply it critically, weighing up where and when it is relevant. For at the heart of each virtue lies the ability to find the right balance between the extremes. In this respect, Parzival is given something more than a mere

---

[9] See for instance the *Introduction to Astrology* by the ninth-century Arabic-era astrologer Abû Ma'sar, which was available in a Latin translation by Johannes Hispalensis. The relevant passage is quoted at length in Klibanski, Saxl, and Panofsky, *Saturn and Melancholy* (New York: 1964), pp. 127–28.

code of behaviour; he is challenged to develop moral imagination, and if at some time he should adhere too strongly to one moral precept while ignoring its opposite, he is likely to go astray. Despite Gurnemanz' heartfelt wish that he should stay and become a kind of son to him, Parzival feels that he has not yet proven himself in the world, and sets out once more in quest of adventure,

He continues on to the city of Pelrapeire, where Queen Condwiramurs is desperately trying to fend off an unwanted suitor, a man who is determined to win her love by force of arms. Parzival leads her wearied troops to victory and wins the hand of the queen in marriage. But after a brief period of conjugal bliss, a great longing takes hold of him to see his mother once more. Not knowing where to look for her, he gives his horse free reign, and is carried further in a single day than a bird could easily fly. Parzival, of course, is ignorant of the fact that his mother Herzeloyde had died of grief when he left her standing forlorn in her doorway, without so much as looking back. But by releasing the reins of his horse, he is led unerringly to the very place with which she is most deeply connected, and where her brother Anfortas has been entrusted with the Grail kingship. As Parzival approaches Munsalvaesch, he sees a fisherman angling in a lake from a small boat. Parzival asks where he might find lodging for the night, and is given directions to a castle, where the fisherman himself promises to be his host.

In contrast to the story as related by Chrétien, the tragic grandeur of the events which unfold during Parzival's first visit to the Grail Castle cannot be fully understood without considering their astronomical background, above all Saturn's baleful influence. Wolfram criticises Chrétien for these omissions, contrasting his account with that passed on by Kyôt, whom Wolfram cites as his source.[10] To be sure, in describing Parzival's first visit to the Grail Castle, Wolfram makes no explicit mention of any cosmic forces at work; he waits until his hero, many years later, arrives at the hermitage of Trevrezent, who reveals to him the macrocosmic dimension of what he had experienced. Parzival learns that at the very moment of his arrival at the castle, Saturn had returned to its goal or *zil,* making its influence stronger than at any other time since Anfortas, the Grail King, had been wounded. In Trevrezent's words, 'That the planet Saturn had returned to its point of departure, was revealed to us by the wound, and by the summer snowfall'.[11] The severe cold brought on by the planet

---

[10] 'ob von Troys meister Cristjân disem maere hât unreht getân, daz mac wol zürnen Kyôt, der uns diu rehten maere enbôt', ('in doing an injustice to this story, Master Chrétien may well have aroused the wrath of Kyôt, to whom we owe the true version', *Parzivâl*, pp. 827, 1–4. On the possible identity of Kyôt, see footnote 25.

[11] 'dô der sterne Sâturnus wider an sîn zil gestuont, daz wart uns bî der wunden kuont, unt bî dem

greatly increased the pain of the king's still festering wound, and a day or two later sent an untimely snowfall.

Not only Saturn, but a number of planetary 'returns', as well as the New and Full Moon, can intensify the king's agony.[12] Later we learn that a conjunction of the two other superior planets, Mars and Jupiter, can have a powerful effect: 'Time had moved forward, and Mars and Jupiter had returned with wrathful pace ... to the place from which their revolutions had begun. He (Anfortas) was defenceless, his wound caused him renewed anguish'.[13]

Just what Wolfram may have meant by the word *zil* has long been a subject of debate. Literally, *zil* means 'goal', while Wolfram specifies that it signifies the return to a starting point. The *zil* of a planet could be understood to mean its domicile in the zodiac, the place where, according to astrological tradition, it could be expected to develop its greatest power and influence.[14] While this interpretation has much to recommend it, the word *zil* is not known to have been used elsewhere as a synonym for domicile, which is consistently called *hûs*, or house, in medieval German astrological sources. Besides, Wolfram stresses that, when reaching its *zil*, 'Saturn rose up so high'[15], while Capricorn and Aquarius, its daytime and night-time domiciles, are among the nethermost signs of the zodiac. An alternative possibility, that the planet's exaltation is meant, encounters similar difficulties. The exaltation of a planet was considered to be the specific point in the zodiac where it unfolds its greatest power.[16] It is located opposite the planet's 'dejection' (*deiectio)*, where its influence is weakest. According to Ptolemy, the exaltation of Saturn lies in the sign of Libra.[17] The sign of Libra, however, does not rise higher in the sky than the Sun at the time of the autumn equinox, and Saturn in this position is also descending in its revo-

---

sumerlîchen snê'. *Parzivâl*, p. 489, 24–27.

[12] 'etslîcher sterne komende tage die diet dâ lêret jâmers klage, die sô hôhe ob ein ander stênt und ungelîche wider gênt: und des mânen wandelkêre schadet ouch zer wunden sêre'. *Parzivâl*, p. 490, 3–8.

[13] 'nu hete diu wîle des erbiten, daz Mars unde Jupiter wâren komen wider her al zornec mit ir loufte (sô was er der verkoufte) dar si sich von sprunge huoben ê. Daz tet an sîner wunden wê'. *Parzivâl*, p. 789, 5–10.

[14] Wilhelm Deinert, in his excellent study 'Ritter und Kosmos in Parzival', takes this view, defending it against other interpretations (pp. 24–29).

[15] 'Sâturnus louft sô hôhe enbor', *Parzivâl*, p. 493, 1.

[16] 'For the other planets, one can only speculate as to why their exaltation or dejection should lie in one or another sign of the zodiac. Because Saturn is the coldest planet, it has its exaltation where the sun has its dejection, i.e., in Libra. ...Here too Saturn can be regarded as a kind of nocturnal anti-sun'. (Boll, Bezold, Gundel, *Sternglaube und Sterndeutung*, p. 59 my translation).

[17] 'Saturn again, in order to have a position opposite to the sun, took ... Libra as his exaltion'. *Tetrabiblos*, I, 19.

lution through the zodiac (i.e., reaching a lower culmination from year to year). It would therefore seem to contradict the description of Saturn 'rising up so high' when reaching its *zil*. A third possibility is that the planet's 'apogee' is meant. Martianus Capella describes the apogee as the 'point at which its orbit reaches its highest elevation above the earth', by which is meant not the highest culmination in the sky,[18] but the greatest orbital distance from the earth. Martianus places the apogee of Saturn in Scorpio.[19] When Saturn passes through this sign it describes an arc only slightly higher than that of the Sun at midwinter. Theoretically, Wolfram may have used the expression 'rose up so high' with relation to the planet's distance from the earth, but whether he possessed the necessary knowledge of astronomical theory must remain an open question.

Let us suppose, on the other hand, that Wolfram was not speaking as an astrologer, but as an observer of the sky. If we give him the benefit of the doubt, and accept hypothetically that he may have been translating or adapting from a source in which Saturn plays a prominent rôle, we should hardly be surprised if he had made some attempt to observe the planet. During the first decade of the 13th century, while he was composing *Parzivâl*, he could have observed Saturn rising higher and higher each year in the zodiac. From 1204 to 1207 Saturn passed between the Hyades and the Pleiades through constellation Taurus into Gemini. At each successive opposition to the Sun it stood higher in the sky, before reaching its highest culmination near the point of summer solstice in the winter of 1208. The greatest brilliance of the planet always took place during the long winter nights, when it rose near sunset, culminated at midnight and set around sunrise. The immediate experience of the longer visibility and higher ascent of the planet could quite naturally have led him to the view that any planet would attain its greatest power and influence at such times. In stressing that the planet, upon reaching its 'goal', is returning to the position from which its revolution began, Wolfram emphasises the cyclic nature of its movements. Saturn's majestically slow sidereal revolution of 29½ years would certainly ensure that its return to its *zil* is the first since Anfortas received the wound.

The possibility that Saturn rising 'up so high' might be pointing to a high culmination, like that of the midsummer sun, has, to my knowledge, never been seriously considered in academic circles. But it makes sense if Wolfram was relying on visual astronomy. Let us suppose that Wolfram, as a knight and a singer, was not deeply versed in astronomy and astrology, but nevertheless convinced that the astrological imagery in his source was essential to an understanding of the story. If this were the

---

[18] Martianus Capella, *The Marriage of Philology and Mercury*, ¶ 884.
[19] Capella, *The Marriage of Philology and Mercury*, ¶ 886.

case, it is quite possible that he would use German expressions which were not usual in astrological texts, and that he might inadvertently give his own, idiosyncratic interpretation, based on his observations. If this were, indeed, the case, it would bear out the view given in the *War of the Minstrels,* insofar as it would rule out Wolfram having had any formal schooling in astrology.

We have yet to consider the tragedy of Parzival's failure to ask the question which might have healed Anfortas and brought redemption to the Knights of the Grail. Parzival cannot be forewarned, for the question must arise spontaneously from the compassion of his heart. The asking of the question at the most dramatic moment of the Grail ceremony was obviously regarded as something momentous, and its omission as a sin which could scarcely be forgiven. In Parzivâl's case, the reader is assured that in the very moment when he would like to have asked the question, he thought of the advice given him by his mentor: 'Gurnemanz advised me – and he surely had my well-being in mind – not to ask so many questions.'[20]

The description of Parzival's arrival at the castle seems to offer a hint that he may not be inwardly prepared for what he is to experience there. He is warmly welcomed, and after surrendering his armour and weapons, is led to his room. There he is given a magnificent cloak of rich Arabian cloth and is told that the Queen, Repanse de Schoye, lends it to him, for his own clothing has not yet been cut. There seems to be an intimation here that Parzival has been expected, but not so soon. He offers his most gracious thanks for the honour of wearing the Queen's cloak, but now, whether by plan or by accident, he is put to a test which almost proves too much for him. A man with a glib tongue, possibly a sort of court fool, tells him that the king is angry with him and demands to see him at once. Parzival, not recognising this remark for the jest that it is, looks about for his sword, and luckily not finding it, he balls his hands into fists with such violence that blood spurts forth from beneath his fingernails, running down to wet his sleeves. Although it is not expressly stated, we may well imagine his blood staining the sleeves of the Queen's cloak, which he has just been given. Outwardly, little is made of this, and Parzival is led into the hall, where the Grail procession and the meal dispensed by the Grail will take place before his astonished gaze. But as a prelude to these rites, a young page enters the hall, carrying a lance aloft as he walks slowly past all the four walls. Drops of blood trickle down the shaft from the steel tip, until they are arrested by his sleeves; all the knights who are present wail and weep in their sorrow, and their lamentations only stop when the page once again leaves the hall. Here we are clearly confronted with an image of a garment being stained by blood,

---

[20] *Parzivâl,* pp. 239, 11–13.

and the similarity to the image preceding it is most disturbing. As Parzival later learns from Trevrezent, this was the very lance which had wounded Anfortas, and to relieve the intensity of his pain, the knights had learned to lay its tip against his wound, allowing it to draw out the extreme Saturnine cold, causing ice crystals to form on the tip of the spear. On this occasion, however, the pain was so intense that this remedy proved of no avail, and the spear had to be thrust deeply into the wound before it could bring relief. Understandably, the knights were bewailing the suffering of their king, but they might equally well have bewailed the unreadiness of their guest to ask the question arising out of the compassion which was expected of him. Indeed, he had arrived too soon – if not too soon for Anfortas, then too soon with respect to his own biographical and spiritual development.

The failure of Parzival to ask the question may seem a simple enough failing, and we might imagine that a few weeks longer under the tutelage of Gurnemanz might have sufficed to show him that there are moments when a spontaneous expression of empathy is more important than all the rules of courtly conduct. But the failure to ask the question need not necessarily point to a lack of feeling. In a later French version of the Grail legend, *Perceval le Gallois ou le conte du Graal*,[21] we read of the failure of another knight, Gawan, to ask the question, even though he was expressly warned that he must do so. He sees 'two damsels that issue forth from a chapel, whereof the one holdeth in her hands the most Holy Graal, and the other the lance, whereof the point bleedeth thereinto.'[22] Gawan falls into so deep a reverie at this sight that he is quite unable to speak. The damsels pass into another chapel, while Gawan sits deep in thought, to the great consternation of all those present. They come forth again, and this time Gawan's gaze is arrested by three drops of blood which fall upon the white tablecloth. Once again he is unable to speak, and when he attempts to kiss the three drops of blood, they vanish.

It would seem certain that the failure, or rather inability of Gawan to ask the question was not due to a lack of feeling or empathy. If anything, he was so overwhelmed by reverence and awe that he was quite unable to escape from a purely passive form of participation. Indeed, as a pupil of the ancient mysteries, he would have been exemplary. The neophyte was expected to follow obediently whatever he was instructed to do, his role being essentially passive. What was new and absolutely unique in Parzival's initiation, if we may so call it, is the requirement that he seize the initiative, that he cast aside all the rules of conduct he has learned, and speak

---

[21] Translated into English as *The High History of the Holy Graal* by Sebastian Evans, (Cambridge & London, 1969), pp. 88–89.

[22] *The High History*, p. 88.

what has never before been spoken. Only so can the Grail King be healed, only so can Parsifal become worthy to wear the cloak of the Grail Queen.

What then was the cause of his failure? It is obvious from Wolfram's narrative that he had not fallen into a trance like that of Gawan at the sight of the three drops of blood on the tablecloth. Nor is he lacking in sensitivity. The question, we are told, is nearly asked, and the words which should have been spoken outwardly were tentatively spoken inwardly in his own mind. But in the critical moment he remembered Gurnemanz' admonition not to ask too many questions. Failing to achieve the moral imagination suited to this unique moment, he remains unfree, and his silence appals all those assembled there, while he resolves to postpone the question until the morrow.

When he wakes the next morning to find himself alone, with no one to help him dress or put on his armour, he once again becomes angry. But it was not anger, nor was it any other obvious vice which had held him back from asking the question. The fatal influence of the moment would rather appear to be the paralysing effect of Saturn, bringing its chill to the wound, inhibiting Parzival from speaking, and after a day or so, blanketing the landscape with snow. Parzival had arrived at the wrong cosmic moment; he was not expected so soon, and his robes had not yet been cut. He was not yet equal to the pall of frost and melancholy imposed by Saturn.

After some brief, but significant adventures following his departure from the Grail Castle, Parzival spends a night nearly freezing in the snow, quite unaware that he has strayed close to King Arthur's present encampment on the river Plimizoel. At the first light of dawn, he observes how a falcon, having escaped from Arthur's huntsmen, strikes a goose, from which three drops of blood fall onto the snow before him. Parzival watches, transfixed, as the drops of blood begin to spread out in the white snow, and the delicate tints of rose remind him of his beloved wife Condwiramurs. So powerful does his yearning become that he loses himself altogether, as his gaze becomes a fixed stare. The similarity to the image described above of the three drops of blood on the tablecloth is striking. Just as Gawan is overcome by his love for and devotion to his Saviour, and thus rendered incapable of taking the initiative expected of him, so Parzival is transfixed by the sight of the three drops of blood in the snow. Wolfram assures us that he is quite in the power of Frau Minne, Lady Love. But we must not forget that all of nature has now fallen under the spell of Saturn, the planet which has brought the snowfall, and on the human level, a kind of paralysis of the will, a tendency to surrender to the spell of Saturnine inwardness. Parzival loses himself in the memories which rise up in him; the past becomes more real than the present, and to all appearances he has fallen into a hypnotic state. To be sure, he may be a captive of Frau Minne, of Venus, but it is her father Saturn who has delivered him into her hands.[23] As he gazes down from his horse to the image which holds him captive, chance would have it that his lance is pointed upwards, a sign which another knight can only interpret as a challenge to fight. One knight after another from King Arthur's court attempts to throw him from his horse, and they are only prevented from doing so when Parzival's own steed turns about in alarm, so that he can no longer see the drops of blood. Momentarily awakening, he easily knocks them down onto the snowy ground, only to return at once to his reverie. At last Gawan, who has learned from his own experiences with Frau Minne, recognises Parzival's plight for what it is, and releases him from his stupor by throwing a piece of cloth over the drops of blood. Returning to a normal state of consciousness, Parzival expresses his gratitude; the two knights vow friendship and ride together to the king's encampment.

We have now reached an important threshold in Parzival's biography. At first it would appear natural that he should join the Arthurian court, where he has long been awaited, and now that he finally arrives, he is welcomed with jubilation. But a new catastrophe awaits him when a most remarkable lady from the Orient arrives, bearing sorrowful tidings. Her name is Cundrie, and she is called 'La Sorcière'. She speaks Arabic, Latin and French, and is learned in the seven liberal arts, including, Wolfram stresses, astronomy. Few jousts were ever fought for this damsel's love, for her nose is like that of a dog, and great tusks like those of a boar protrude from her jaws[24]. We need not elaborate her numerous other beauty marks, which are much of a character with these. She addresses the king in French, warning him of the shame and perfidy that will fall on his court should Parzival join it. Then turning to Parzival himself, she confronts him publicly with his lack of compassion for the sorrowful Fisher King, saying that his heart is void of feeling, and giving him little hope of ever atoning for his failure. In the midst of the dismay and confusion which follow in the wake of Cundrie's words, yet another messenger arrives at the court, falsely accusing Gawan of slaying his lord in the act of greeting, and challenging him to single combat in forty days' time. Thus it chances that a shadow has fallen on the reputations of Parzival and Gawan alike. The accusation levelled against Gawan is manifestly

---

[23] The hermit Trevrezent later explains that the unseasonal snow of this night had been brought on by Saturn: 'Saturn was moving high in its orbit, the wound (of Anfortas) seemed to feel it in advance, before the outer frost arrived. The snow only fell the following night'. ('Sâturnus louft sô hôhe enbor, daz ez diu wunde wesse vor, ê der ander frost koem her nâch. Dem snê was ninder als gâch, er viel alrêrst an dr andern naht.') *Parzivâl*, p. 493, 1–5.

[24] *Parzivâl*, p. 313, 1–28.

untrue, while Parzival must himself admit that Cundrie's words are all too true. Yet he cannot help feeling that God has done him an injustice, and he does not hesitate to declare his anger against God for so treating him. He is no longer the simple, innocent, yet often foolish young man, described in the early adventures of the story, but doubt has taken hold of him, and will be his companion until such time as he can relieve Anfortas of his suffering. He will have no rest until he has done so, however great the trials he will have to face. His anger against God is, of course a projection, and while it may give him the strength which is born of desperation, it will certainly not lead him to the Grail. That he should so expressly declare his anger against God shows that he must still struggle to transform his own lower nature. And so, both knights feel impelled to leave Arthur's court, Gawan to clear his name of the charge of murder, and Parzival on a quest so lonely that not even Wolfram's audience is allowed to accompany him.

The following episodes are largely devoted to the adventures which Gawan encounters on his way to keep his appointment. All of these are adventures of the heart, involving him in successive amorous exploits, and culminating in a love so powerful that he remains faithful to his chosen lady through the most appalling adversities. During all these episodes, where Parzival is only occasionally glimpsed, as it were, from a discrete distance, Gawan has fully entered the service of Frau Minne, and whatever martial exploits he engages in, they are all done for the sake of love. Parzival does indeed appear from time to time in the background, and report of the extraordinary deeds of a Red Knight occasionally reaches Gawan's ears. The soul-realms which these two knights – perhaps it is not too bold if we call them the Knight of Mars and the Knight of Venus - now inhabit, seem to mirror each other in many ways. Thus, when Gawan asks his hosts what adventure might await him in the enchanted Chastel Merveille, they greatly lament his question: while Parzival causes sorrow by not asking, Gawan has quite the opposite experience. He nevertheless succeeds in breaking the enchantment laid upon the castle by the absent magician Clinschor, and frees the knights and ladies who had been held captive there.

In the ninth Book Wolfram interrupts his narrative of Gawan to return to Parzival, just before he arrives at the solitary home of his maternal uncle Trevrezent, the brother of Anfortas. The moment has come, he tells us, to explain why Chrétien de Troyes did not do full justice to the story. Trevrezent, as we have already learned, will help Parzival understand the celestial events which were synchronous with his own experiences at the Grail Castle. It is therefore obviously a fitting moment for Wolfram to divulge his own source and to offer some justification for the cosmic dimensions of the story.

Wolfram assures us that he has translated his own story from the French of a certain Kyôt, the famous Provencal singer[25]. Kyôt seems, however, to have been at least as much a scholar as a singer, for he succeeded in unearthing an Arabic manuscript lying abandoned in Toledo, in which he discovered the first written account of the Grail. Kyôt's knowledge of occult signs and sigils, together with his baptism, enabled him to understand it, even better perhaps, than had its own author. He names the heathen author as Flegetanis, and calls him a *fisîôn*, a word not known to occur elsewhere, and which has variously been rendered as 'physicus', 'physiologist' and 'physiosophist'[26], all of which are intended to mean something like a natural scientist. It might, however, equally well be understood to mean a 'visionary'. Flegetanis was a heathen on his father's side, and on his mother's a descendent of Solomon, so that he can be thought to have united within himself the wisdom and learning of Islamic and Hebrew tradition. Wolfram says of him:

> The heathen Flegetanis could describe each planet's departure and future return: how long each circles in its orbit before it once more reaches its goal. The revolutions of the planets shape human character. Flegetanis the heathen saw hidden mysteries, of which he spoke with great reticence. He said, 'There is a thing called the Grail'. Its name he read clearly in the stars. 'An (angelic) host brought it to the earth; and then flew higher than the stars, assuming that their innocence allowed them to do so. Since then it has been entrusted to baptised Christians, who maintain the purest chastity. All those called to serve the Grail are of the highest worth and merit'.[27]

---

[25] 'Kyôt ist ein Provenzâl, der dise aventiur von Parzivâl heidensch geschriben sach, swaz er en franzoys dâ von gesprach…' ('Kyôt was a Provencale, who found these adventures of Parzivâl written in Arabic, which he rendered into French'), *Parzivâl*, p. 416, 25–28; There was, in fact, a trouvère known as Guiot de Provins (a town in Champagne), though nothing but the outward similarity of the name could possibly suggest that he might be identical with Wolfram's Kyot. Most scholars now assume that Kyot and Flegetanis were both literary devices of Wolfram, which he created to justify his departures from Chrétien's original. If Kyot had been a 'famous Provencal singer', a troubadour, it seems unlikely that his name would be so completely forgotten today. If, however, he had been primarily a scholar and translator of Arabic astrological and mystical works, he might be less likely to have achieved lasting fame. At the very least, whether he was a real historical personality or a fiction of Wolframs, Kyot, like Parzival's father Gahmuret, can be taken as a representative of the meeting of east and west, of the mutual fructification of cultures. And this was certainly a concern close to Wolfram's heart.

[26] In the English translation by A.T. Hatto (physicus) and the German translations by Peter Knecht (physiologist) and Dieter Kühn (physiosophist), all referring to *Parzivâl*, p. 453, 23.

[27] *Parzivâl*, p. 454, 9–30.

Kyot, we are told, now began searching in the Latin chronicles of various Christian kingdoms, in the hope of bringing to light which family had been given the custody of the Grail. In Anjou at last, he discovered the ancestral line of the Grail kingship, which was passed on from Titurel to his son Frimutel, and thence to Anfortas. As we soon learn from Trevrezent, Parzival's mother Herzeloyde was Anfortas' sister, making Parzival a nephew of the king. Despite appearances, the kingship was not hereditary. The new king was called to serve the Grail by means of a radiant script which appeared on its surface, and then disappeared again once they had been read and understood.

Further, we learn that Anfortas became king in the flower of his youth, after the untimely death of his father Frimutel. Disregarding the rule that his Queen should be chosen and her identity revealed by the Grail, he set off on his own, in quest of adventure and romance. Having entered the service of an entrancingly beautiful maiden, he once encountered a heathen knight who was searching for the Grail. What an extraordinary situation! The Christian Grail King rides forth seeking glory in the service of a damsel who, as he must know himself, is not destined to become his wife. And he finds himself facing a heathen knight, who has gone forth in search of the Grail! The poisoned spear of his opponent wounds Anfortas in the testicles,

Flegetânis der heiden
kund uns wol bescheiden
ieslîches sternen hinganc
und sîner künfte widerwanc:
wie lange ieslîcher umbe gêt,
ê er wider an sîn zil gestêt.
mit der sternen umbereise vart
ist geprüevet aller menschlîch art.
Flegetânis der heiden sach,
dâ von er blûweclîche sprach,
im gestirn mit sînen ougen
verholenbaeriu tougen.
er jach, ez hiez ein dinc der grâl:
des namen las er sunder twâl
inme gestirne, wie der hiez.
,ein schar in ûf der erden liez:
die fuor ûf über die sterne hôch,
op die ir unschult wider zôch.
sît muoz sîn pflegn getouftiu fruht
mit alsô kiuschlîcher zuht:
diu menscheit ist immer wert,
der zuo dem grâle wirt gegert.'

making him unfit for the kingship, and plunging the whole community of the Grail into deepest sorrow.

These revelations, both regarding the source of the story, which we learn from Wolfram directly, and regarding the history of Anfortas' wound, which we learn from Trevrezent, raise far-reaching questions. The very existence of the Grail was discovered by a heathen astrologer-philosopher. He was able to read its name clearly in the stars, and to describe how it had been brought to the earth by a heavenly host, by those angels who had remained neutral when Lucifer rebelled against his Creator. Moreover, we hear of a heathen knight who had ridden out in search of the Grail. In the very moment when the Christian Grail King lapsed from the task entrusted to him, their meeting took place and the tragedy occurred. Are we then to understand that the Grail, in its macrocosmic nature, could best be understood by a visionary pagan astrologer, and microcosmically, belonged in the domain of Christianity?

In the thirteenth-century, the knowledge and practice of astronomy and astrology were far more advanced in the Islamic world than in Christendom. But how would knowledge of astronomy and astrology lead a heathen scholar to discover what has become one of the central mysteries of Christianity? It is surely one of the most re-markable features of Wolfram's method, that he raises questions which he does not directly answer, but rather leaves open for his listeners to live with. His description of the relationship between Flegetanis and Kyot points clearly to a development in consciousness taking place in the course of human history. The heathen visionary philosopher is aware of the macrocosmic mysteries of the Grail. His ability to read the name of the Grail in the stellar script clearly points to a clairvoyant faculty far transcending what he might have learned through any intellectual interpretation of the planetary rhythms. From the description Wolfram gives, I believe it is clear that in reading the stellar script, he is not consulting an ephemeris or applying traditional astrological principles to interpret the planetary movements, but that he is reporting an actual experience. He is able to describe how the Grail was brought to the earth, but from this point onwards, the tale can only be continued by a Christian, whether we call him Kyot or Wolfram. For whatever occurs on the earthly plane requires an internalisation of the cosmic forces of which Flegetanis was able to speak. If the qualities and virtues of the planets are not made a part of the human soul, if they are not purified and transformed into inner faculties, they can still impose their will from without. Saturn would never have been able to torment Anfortas as it did, had he developed the virtues of Saturn within himself. Nor would it have inhibited Parzival from asking the question. And what were these virtues?

In much of medieval tradition, Saturn's influence was regarded much as it is described by Trevrezent. In Alain de Lille's *Anticlaudian* there is an impressive account of Prudentia's journey through the sphere of Saturn. The description might almost have been taken directly from *Parzivâl*: 'She shivers from the frost and snow of the winter solstice and is astounded by the paralysing cold amidst the heat. There winter rages, summer rigidifies and the very heat grows frigid... Pain and distress hold sway, tears, discord, terror, grief, fear, lamentation and injustice.'[28] But there is another side to Saturn which Alain places alongside these negative qualities: 'And yet his tone does not stray from the (celestial) choir, but rather surpasses the voices of his companions... to which he imparts more vitality through the sweetness of his voice.'[29] In Neoplatonic tradition, Saturn was regarded almost exclusively in a positive light. It was seen as the most exalted of the planets, imparting the gift of intellect (*Nous*) and the power of contemplation.[30] It was essentially a Neoplatonic idea that the planets were not evil in themselves, but might appear to work evil if the vessel receiving their influences were impure. In a sermon attributed to Meister Eckhart we read that the 'cleanser' or purifier among the planets was none other than Saturn: 'So in the heaven of the soul,' he writes, 'Saturn is a cleanser giving angelic purity and bringing about a vision of the godhead: as our Lord said, 'blessed are the pure in heart, for they shall see God'.'[31] As the planet of deep contemplative inwardness and purification, we may imagine Saturn as initiating the process of interiorisation, through which the human soul becomes a 'spiritual heaven', adorned with spiritual stars. During the four and a half years of Parzival's solitary adventures, we may imagine that such an inner ripening is able to occur. This is, perhaps the reason Wolfram refrains from offering a detailed account of his adventures; the most significant events are of an inner nature, and cannot easily be described.

At the end of these wanderings, Parzival and Gawan are once again re-united at the itinerant court of King Arthur. Riding out alone one day, Parzival encounters his half-brother Feirefiz, whom he has never seen, and who has come with a great fleet of ships from his kingdom in the Orient. The two fail at first to recognise each other, and, as was all too customary in such cases, begin at once to fight. Neither has ever been defeated, but each soon begins to wonder if he has finally met his match. At last Parzival's sword breaks, but Feirefiz refuses to take advantage of his weakness.

---

[28] Alain de Lille, *Anticlaudian*, IV, 8. Quoted in Klibansky, Saxl, and Panofsky, *Saturn and Melancholy*, p. 186.

[29] Klibansky, Saxl, and Panofsky, *Saturn and Melancholy*, p. 186.

[30] Klibansky, Saxl, and Panofsky, *Saturn and Melancholy*, p. 153.

[31] Quoted in *Saturn and Melancholy*, p. 168. In the original: 'Alsô wirt an dem himel der sêle Saturnus (der fürber) der engelischen reinekeit unde bringet zu lône anschouwunge der gotheit...'

They rest, and as they do so, question each other until they discover that they are both sons of Gahmuret. All doubt is finally removed when Feirefiz takes off his helmet, showing that his skin is mottled with black and white patches like a magpie – a characteristic which Parzivâl had learned of from hearsay. Feirefiz is by nature jovial, and frequently calls on his gods Jupiter and Juno. The contrast between the majesty of this oriental king, who rules over many nations, and Parzival, the knight who wanders homeless and alone, ever true to his inner quest, could hardly be greater. Parzival leads his brother to Arthur's tents, where he is made acquainted with the most celebrated knights and ladies of the court. In this moment Cundrie la Sorcière arrives, this time wearing the insignia of the turtle dove, which marks her as a messenger of the Grail. She announces that Parzival's name has appeared on the Grail, and that he has been deemed worthy to return to make good what he omitted to do on his last visit to the Grail Castle. She now outlines the sphere over which Parzival is called to rule:

> She named the seven planets in the Arabian tongue. The noble and wealthy Feirefiz knew them: he sat before her, black and white. She said: 'Parzival, give heed! The highest planet Zval (Zuhal = Saturn), and the swift Almustri (Al-Mushtari = Jupiter), Almaret (Al-Mirrih = Mars) and the brilliant Samsi (Shams = Sun), bestow their blessings on you. The fifth is called Alligafir (Al-Zuhari = Venus), and the sixth Alkiter (Al-Katib = Mercury), and the closest Alkamer (Al-Qamar = Moon). What I say here is no dream: they are the bridle of the firmament, and slow down its movement, ever striving against its course. Your time of sorrows is ending. All that the planets encircle, and all that their light envelops, is marked out as your goal (zil), to attain and to keep...[32]

The Arabian names are given in descending sequence, beginning with the outermost or superior planets, and continuing to those nearest the earth. The idea that the planets in their movements strive against the diurnal motion of the fixed stars can only be understood in the context of the geocentric system of Ptolemy. The entire starry firmament was believed to encircle the earth in the course of a day. The planets, which move from west to east against the background of the fixed stars, complete this motion somewhat more slowly. The planets having the fastest diurnal movement are therefore those whose movement through the zodiac is slowest, namely Saturn and Jupiter. Therefore, Cundrie speaks of 'swift Almustri'. The superior planets Saturn, Jupiter, and Mars are further said to bestow their blessings on Parzival. Is it accidental that just these three should be singled out in this way?

---

[32] *Parzivâl*, p. 782, 1–21.

As Parzival approaches Munsalvaesch in the company of Feirefiz – for he is allowed to take one companion with him – we learn that a new planetary configuration is causing Anfortas' wound to flare up once more, making his pain almost unbearable. The different manuscripts describe this event somewhat differently; in the version most generally used, we read that 'Mars *or* Jupiter had returned all wrathful in their courses ... to the places from which they had set out. Anfortas felt it in his wound'[33] Other manuscripts have either 'Mars Jupiter' or Mars *and* Jupiter. In view of the plural form of 'their courses', it would seem likely that both planets were involved, and that a scribal error is responsible for the use of 'or' instead of 'and'. Arthur Groos, in his excellent study *Romancing the Grail,* proposes that a conjunction of the two planets may have been meant, and points out that conjunctions of Mars and Jupiter were traditionally thought to bring about catastrophes and pestilence.[34] He quotes Albertus Magnus, who wrote: 'For a conjunction of the two particular planets, Jupiter and Mars, ... creates pestilential winds and corrupts the air, which quickly kills many men and animals'.[35] Various Arabian sources concur in this view. Groos cites Abû Ma'shar and Abû al-Hasan 'Ali, who writes, '*Quando iungitur Mars cum Jove significat mortem regis per epidemiam*'.[36] Most of the relevant astrological passages derive ultimately from translations made in mid-twelfth-century Toledo,[37] where Kyot was said to have made his own discovery of the original version of the Grail story.

Cundrie, in listing the planets which 'encircle and envelop' Parzival's future reign, would seem to imply that he has become so far master of their virtues within himself, that he can now be regarded as an ideal microscosmic image of the macrocosm. All through his adventures, he has been known as the Red Knight, and we have been at pains to show that he has had to struggle with the impulsiveness and hot temper of Mars. Once he is called to the Grail, however, he has no doubt learned so much

---

[33]   Parzivâl, p. 789, 5 – 10.

daz Mars oder Jupiter

wâren komen wider her

al zornec mit ir loufte …

dar si sich von sprunge huoben ê.

das tet an sîner wunden wê.

[34] Arthur Groos, *Romancing the Grail: Genre, Science and Quest in Wolfram's Parzival,* pp. 200, f.

[35] Quoted in Groos, *Romancing the Grail,* pp. 201–2, 'Coniunctio einim duarum paecipue stellarum, quae sunt Iuppiter et Mars,… faciunt ventos pestilentes, et aeres corruptos, qui subito necant multitudinem hominum et animalium'. Albertus Magnus, *De causis proprietatum elementorum* II.ii.1.

[36] Groos, *Romancing the Grail,* p. 203.

[37] Groos, *Romancing the Grail,* p. 204.

through suffering that he can hope to provide a true counterweight to the negative influences of the planet. His solitary years, spent under the tutelage of Saturn, have taught him to make the planetary virtues into qualities and ornaments of his soul. Parzival and Feirfiz, as the earthly representatives of Mars and Jupiter, adorned with their highest attributes, approach the Grail Castle just as the two planets overhead return 'wrathfully in their courses'. Together they can heal what has gone amiss in the Grail community: Parzival by his words of compassion, which bring healing to the king, and Feirefiz by his pure and deep love for the queen, Repanse de Schoye, whom he marries after receiving the sacrament of baptism.

Anfortas, when he is healed, appears even more beautiful than Parzival himself, who was often likened to a second sun in his own radiance:

> What the French call 'bloom' – its light came over his skin. Parzival's beauty was as nothing beside him, nor that of Absalom, the son of David, nor of all those whose beauty has been most celebrated, ... none of these was equal in beauty to Anfortas when he left illness behind. God's artistry is still undiminished. (796, 5–16)

It would not seem too much to say that Saturn, as the great purifier among the planets, has so purged and refined Anfortas' soul through years of suffering, that even his physical body now shines like a celestial light. He appears as the New Adam who has passed through sin and repentance to become once more the image of God. The script which now appears on the Grail designates Parzival as the new Grail king, while Anfortas is once again free to serve the Grail as a knight.

In conclusion, let us return once more to the question addressed in the *War of the Minstrels:* was Wolfram an adept in astronomy and astrology, or was he passing on traditions which he only partly understood? Was the devil Nasion right in calling him a contemptible layman (*ein laie schnickeschnack*)? Did the Wolfram depicted in that poem describe himself accurately when he claimed total ignorance of both sciences? His vivid descriptions of astronomical phenomena have long puzzled scholars; the fact that he does not employ the current astrological terminology makes it difficult to be sure what he means. Some suggestions have been offered here, although it may never be possible to demonstrate conclusively what he meant when he spoke of a planet's 'goal'. For this reason it seems likely that he did have a source or authority from which he drew, translating the terminology as best he could into his own imagery and experience. What can be said with some confidence is that he had a deep understanding of the 'interiorisation' process which he describes taking place through the influence of Christianity. The task of the custodians of the Grail, in Wolfram's view, clearly involved what we might describe as an alchemy of the soul, whereby the planetary virtues, once attained, adorn the soul and liberate it from the compulsion imposed by the stars and planets. The question still remains open what celestial phenomena may have led Flegetanis to his discovery of the Grail in the macrocosm, of which he spoke with great reticence and veneration. Microcosmically, the Grail seems nothing less than a symbol of the interiorisation process, the creation of what might be called an inner vessel, capable of receiving the spiritual inspirations which flow from the cosmos.

Seeing the World Soul:
Marsilio Ficino and Talismanic Art

Ruth Clydesdale

ABSTRACT: In his *De Vita Libri Tres*, Marsilio Ficino cautiously describes the methods for inscribing magical figures upon stone and metal. He also regales the reader with a tale of how such talismans can go disastrously wrong. However, he overcomes his hesitation and doubts sufficiently to recommend the creation of an archetypal talisman that would result in a work of art or craftsmanship of considerable complexity. This representation of the cosmos is intended as an object of meditation, and its use signifies the pinnacle of a progression of spiritual exercises linked to the planets. I shall argue that Ficino's known links with practising artists makes it entirely possible that other paintings from fifteenth-century Florence were intended as magical items that could play a role in the planetary magic that Ficino espoused.

> ...in paintings and buildings...the disposition and shape as
> it were of the soul itself is perceived.
> —Ficino, *Theologia Platonica*,
> Book X, Ch.IV

Imagine the marketplace in Florence on a summer evening in 1482. Two friends meet by chance and talk together. One is the painter, sculptor and goldsmith Antonio Pollaiuolo, and his chance-met friend is the philosopher, magus, musician and melancholiac Marsilio Ficino. Pollaiuolo passes on greetings to Ficino from a mutual friend.

This miniature glimpse into Ficino's life, described in one of his *Letters*, demonstrates his intimacy with the Florentine art world.[1] Ficino was friends with both the Pollaiuolo brothers, who ran one of the most successful workshops in Florence. He also counted the architect Leon Battista Alberti amongst his close friends and admirers; Domenico Ghirlandaio painted his portrait and it's thought from links between some of Ficino's writings and Botticelli's paintings that they too were acquainted.

Several scholars have asserted that Ficino had little interest in art, but the friendships listed above suggest otherwise. Edgar Wind's claim that 'Ficino's own visual sensibility was slight' pales in face of the strong tradition of astrological imagery that Ficino inherited and utilised, to say nothing of his use of images in talismanic magic.[2] Both art and astrology are fundamental to Ficino's philosophy and his mag-

---

[1] See *The Letters of Marsilio Ficino*, trans. Language Dept of School of Economic Science (London: Shepheard-Walwyn, 10 vols. to date, 1975-), 6 (1999), p. 40.

[2] Edgar Wind, *Pagan Mysteries* (Oxford: Oxford University Press, 1980), p. 127. On astrological imagery in the fifteenth century, see Jean Seznec, *The Survival of the Pagan Gods* (Princeton: Princeton University Press, 1972), pp. 184–215.

ical practice, and the two subjects are inextricably intertwined.[3] They both represent aspects of the harmonious, balanced and gracious life that was Ficino's ideal means of achieving *mens sana in corpore sano.*

A key text for Ficino's views on both art and astrology is Plato's dialogue *Timaeus*, in which Plato gives an account of creation that includes a description of the destiny of the human soul. *Timaeus* took on a particular resonance for him seen through the twin prisms of astrology and art. He was also influenced by the (possibly spurious) Platonic dialogue *Epinomis* in which astrology is described as the highest art and a theory is elaborated of elemental *daimons* under the control of the planetary deities.[4]

In *Timaeus*, Plato argues that the only true reality is spiritual because, unlike physical reality, it is eternal and unchanging. The Creator of the universe or Demiurge is purely good and beautiful, and he created a cosmos that is 'a single visible living being'.[5] He also made human souls from the same stuff as the soul of the cosmos, and instructed them that each had a home star to which it would return if it lived well on earth. Having received this advice, the souls were immured in physical bodies made by the gods under the Demiurge's command.

Plato explains that because human souls are microcosms of the cosmos, their natural state is one of harmony with the orbits of the planets, which are divine beings.[6] But entry into the material world so confuses the soul's inner orbits that it is necessary to re-align them in order for the soul to fulfil its destiny. The underlying principle of Ficino's philosophy is the achievement of such a harmonious re-alignment, through which to purify the soul and return it to its Creator. He was profoundly influenced by Neoplatonists such as Plotinus, who described his own rapturous experiences of achieving union with the Divine.[7]

Harmony is an aspect of beauty and pleasure, so Ficino's ideal life is one of aesthetic delight as well as moral goodness. Plato's Demiurge created a cosmos that reflects his own goodness and beauty, and appreciating those qualities brings the soul closer to its original state of perfection. The cosmos is, in fact, the ultimate work of art – but we need astrology in order to interpret it.

Timaeus suggested to Ficino that astrology is an absolutely fundamental necessity

---

[3] On the critical debate concerning Ficino and art, see F Ames-Lewis, 'Neoplatonism and the Visual Arts at the time of Marsilio Ficino', *Marsilio Ficino: His Theology, His Philosophy, His Legacy*, eds. Michael J. B. Allen, Valery Rees, Martin Davies (Leiden: Brill, 2001) pp. 326–38.

[4] See Plato, *Philebus and Epinomis*, trans. A. E. Taylor, ed. Raymond Klibansky (London; Thomas Nelson, 1956), 984d–985b.

[5] Plato, *Timaeus*, trans. H.D.P. Lee (London: Penguin, 1965), 30d.

[6] Plato's text marks the first appearance of this belief in Greece. See Introduction to *Timaeus*, p. 9.

[7] See Plotinus, *Enneads*, IV.8.i.

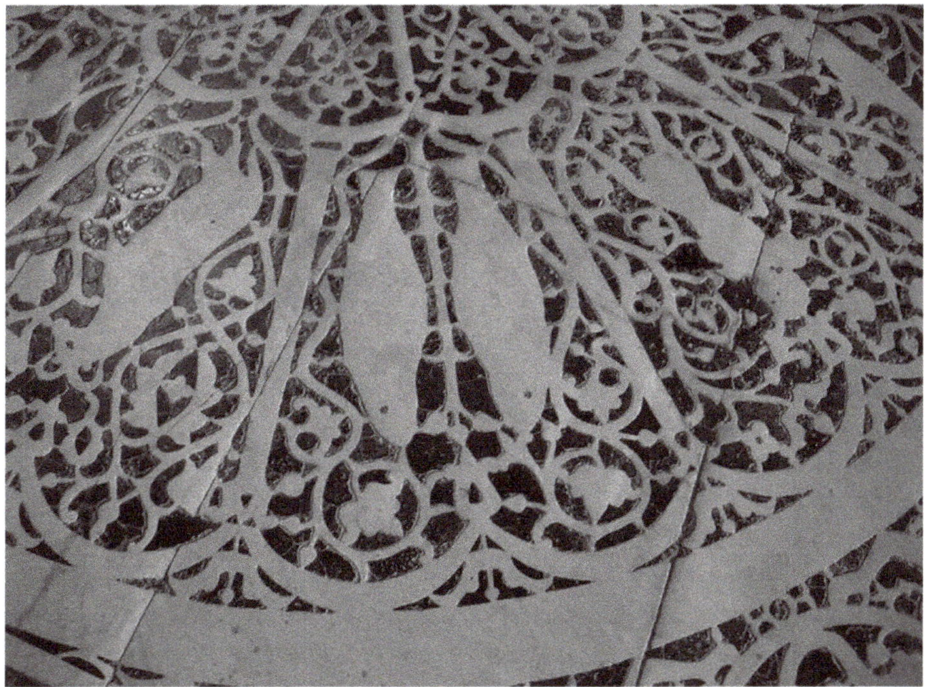

FIGURE 6.1: Detail of zodiac, nave of San Miniato al Monte. ©Sailko

of life. In it he read that the supreme value of the gift of sight is that we can study the orbits of the divine planets, thereby learning to 'use their untroubled course to guide the troubled revolutions in our own understanding...'[8]– that is, unscramble the soul's inner orbits. The whole of Ficino's third *De Vita Libri Tres*, 'De Vita Coelitis Comparanda' or 'On Obtaining Life from the Heavens', is devoted to doing just that, largely by astrological and magical means.

So, astrology and aesthetics are the fundamental means by which fallen souls can return to God by re-establishing themselves in cosmic harmony. Regarding aesthetics, Ficino conferred a philosophico-religious meaning upon the aesthetic that was dominant during his life: the re-creation of classical harmony in art and architecture. The use of number to create harmonious proportion in architecture had already assumed religious significance by the time Ficino was born. A striking example is Brunelleschi's Sagrestia Vecchia in San Lorenzo, the Medici family church. Here, architectural elements in groups of three and four suggest the Trinity, the four Evangelists and (by multiplication) the twelve Apostles.[9] To astrologers, of course, other

---

[8] *Timaeus*, 47a.

[9] See Frederick Hartt, *History of Italian Renaissance Art* (London: Thames and Hudson, 1987), p.148.

meanings also suggest themselves: the qualities, elements and signs for example, or the numbers governing the aspects.

Astrology is, in fact, integrated into the church architecture of Florence. The octagonal Baptistery contains a thirteenth-century marble inlaid zodiac on the floor. Each child entered upon Christian life not only under the gaze of Christ, depicted in mosaic on the Baptistery ceiling, but also in the presence of a symbol of the complete cosmic circle of existence.

Prominently placed on one of the hills surrounding Florence is the church of San Miniato al Monte. This eleventh-century church underwent extensive restoration and rebuilding over seven years from 1200, at which time another marble zodiac was laid down just inside the main doors (FIGURE 6.1). Astrological symbolism is thus enshrined in the two oldest sacred buildings in the city and its environs.[10]

Florence was considered to fall under the first zodiacal sign, Aries, and the importance to the city of its zodiacal sign is indicated by the symbolic significance of a statue of a lion, known as the Marzocco.[11] The lion is the totemic animal of Florence. Astrologically the lion is associated with the sign of Leo. However, the name of the Marzocco is thought to derive from that of Mars, since the lion was commissioned to replace an ancient Roman statue of Mars, which had been washed away from the banks of the Arno during a flood in 1333. The Marzocco that exists today is by Donatello and was sculpted around 1420, a mere decade or so before Ficino's birth. The original is now preserved in the Bargello museum and a copy stands outside the Palazzo Vecchio. The Florentines believed that before the adoption of Christianity their city had been dedicated to Mars. The Baptistery was thought to have been built on the site of a temple of Mars, thus linking in through Aries' rulership of the head with the stories of John the Baptist and San Miniato (who were both beheaded).

This symbol and the two ancient churches are vital parts of the city's identity, suggesting that astrological symbolism lies deep in the very heart of medieval and Renaissance Florence. Indeed, virtually all the major monuments of the city testify to this fact. Giotto's campanile is decorated with bas-reliefs sculpted by Andrea Pisano in the mid-fourteenth century, including one of an astrologer at work casting a horoscope. Also shown are personifications of the lights and planets, and since the campanile is part of the cathedral of Santa Maria del Fiore some of them have been

---

[10] For both zodiacal floors, see Fred Gettings, *The Secret Zodiac: The Hidden Art in Mediaeval Astrology* (London: Arkana, 1987), pp. 164–65, 168–69.

[11] See Janet Cox-Rearick, *Dynasty and Destiny: Pontormo, Leo X and the Two Cosimos* (Princeton: Princeton University Press, 1984), pp. 164–65 for the horoscope of Florence showing Sun, Moon, and Ascendant in Aries.

suitably Christianised. Jupiter, for example, is shown as a portly monk holding a crucifix in his left hand and a generous goblet or chalice of wine in his right.[12]

These bas-reliefs are in a crucial public area next to the great cathedral that embodies both the religious life of the city and also its profound civic pride. They are designed to be easily understood; their style is simple and direct. A passer-by would not need to have Ficino's sophisticated level of education to recognise the planets, and they would be part of the everyday visual experience of the majority of Florentines.

Nearby is Santa Maria Novella. In this great Dominican church is a chapel built during 1345–1355 for Buonamico de'Guidalotti. In the sixteenth century it became known by its present name of the Capelloni degli Spagnoli when it was granted to Eleonora di Toledo, wife of Duke Cosimo I. Included in the ambitious frescos painted by Andrea di Buonaiuto in 1365–67 are female personifications of the seven liberal arts making up the quadrivium and trivium. Each of these figures is seated on a throne decorated with a grisaille bust of the particular planet under which her activity falls. Logic, for example, is seated beneath Mercury, shown as a man absorbed in writing.[13] The planets rule over the entire world of knowledge, the sum total of human wisdom. They represent a vital and vivid code for transmitting information visually, a kind of symbolic shorthand for the eye.

The tradition of using astrological symbolism to decorate sacred architecture is venerable and durable, and it continued into Ficino's own lifetime. A little southeast of the campanile stands Santa Croce. Attached to the main building is Brunelleschi's exquisite little Pazzi chapel, begun in 1433 and built for the family that gave its name to the conspiracy against Lorenzo de' Medici in which Ficino himself was implicated by association. There, inside the dome over the altar (finished in 1459) is a fresco symbolically representing the sky above Florence at one particular moment. It is a horoscope in paint, although unfortunately it is now in such a ruinous condition that it is no longer possible to decipher it.

Returning to the Sagrestia Vecchia in San Lorenzo: in the cupoletta of the sacristy is a comparable fresco that seems to have been commissioned by Ficino's own patron, Cosimo il Vecchio (FIGURE 6.2). This fresco was restored during the 1980s, but even so it is not possible to prove conclusively just what date is represented there.[14] However, in each case there must have been some compelling reason for commem-

---

[12] For the Christianised Jupiter, see Seznec, *Pagan Gods*, p. 161.

[13] P. Stefano Orlandi O.P., *Historical-Artistic Guide of Santa Maria Novella and her Monumental Cloisters* (Florence: Edition S. Becocci, no date), p. 64.

[14] For both horoscopes, see Guiseppe Forti et al., 'Un planetario del XV secolo', *L'astronomia* 62 (Jan 1987): pp. 5–14.

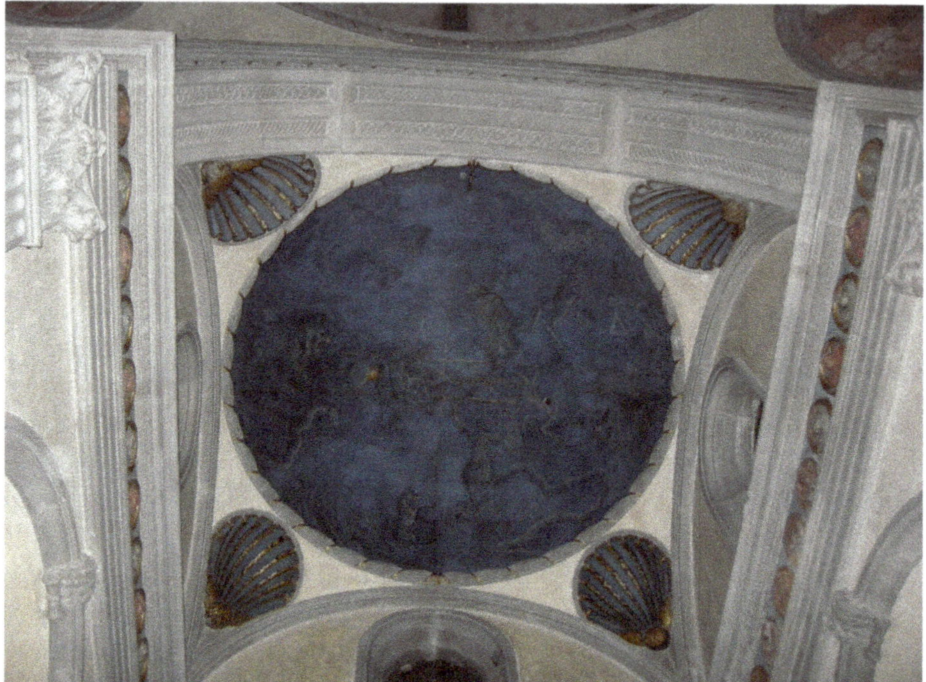

Figure 6.2: Fresco attributed to Giuliano d'Arrigo, c.1440s, Sagrestia Vecchia, San Lorenzo

orating a significant moment in astrological terms.

These examples prove how tightly the cosmic symbolism of astrology was woven into the physical fabric of the Florentine church, just as the Church itself was an essential aspect of fifteenth-century life. Equally, the Marzocco shows that astrological symbolism was just as essential to the city government, that other focus of power in Renaissance Italy. Astrology was used to foster a sense of civic pride and unity; a citizen catching sight of the Marzocco would see his city as a noble place well able to defend itself. In sacred places, representations of the zodiac or planets would serve to remind the congregation of the cosmic order and their place in it, assuring them that the seeming chaos of life has an order and purpose under God. The fresco of the liberal arts in Santa Maria Novella brings astrology into the realm of the cultivation of knowledge upon which civilization is based. And the wealthy could choose to commemorate an important event – perhaps the dedication of an altar, perhaps a personal achievement – in horoscopic form.

All this serves to demonstrate how integral a part of life astrological symbolism was for a man such as Ficino. Essentially, what Ficino did was to bring into the private sphere of his wealthy and educated patrons the visual language with which they were already familiar. Public architectural decoration showed the patron's civic engagement, whereas talismans were for individual use and paintings for private delectation with or without select companions. The astrological symbolism in works of art consequently became more subtle and allusive. As Edgar Wind has observed, patrons steeped in the Neoplatonism that Ficino unfolded would have enjoyed penetrating to secret and personal meanings in paintings.[15]

In the field of painting, Alberti published his influential treatise *De Pictura* in 1435 (Italian version 1436), in which he informed artists that their aim was to produce pleasure by the creation of grace, and their means were the use of mathematical proportion and one-point perspective. Grace was personified in the common Renaissance symbol of harmony, the classical image of the Three Graces. Ficino links these with three benefic planets: the Sun, Venus and Jupiter.[16] His favourite is Jupiter, the most temperate planet and as such the mediator between the other two, and the favours of all three planets are most easily received through the Moon and Mercury. Ficino's astrological instructions for achieving harmony are complex and precise, but his basic rule is very simple. Opposing energies are used to reach the golden mean: 'when you fear Mars, set Venus opposite. When you fear Saturn, use Jupiter'.[17]

Ficino links the Graces with the three types of spirit: natural, vital and animal. Venus rules the natural and procreative spirit, which is absorbed with food and refined into vital spirit, ruled by the Sun, in the heart. Jupiter rules the finest of the three, animal spirit, which is refined in the brain and creates the five senses and thought, including (ideally) an urge towards philosophy and religion.[18] However, the Graces have a deeper Neoplatonic meaning as well. They represent the flow of life from God and back again, and life's final perfection in the spiritual realm. Ficino named the Graces Pulchritudo, Amor and Voluptas – Beauty, Love and Pleasure – and as such they appear on a portrait medal of Ficino's friend Pico della Mirandola, possibly with talismanic intent.[19]

Ficino describes various ways of attracting the benefic power of the Three Graces. His methods are based on the fifth-century Neoplatonist Proclus's description of theurgy in his short treatise *De Sacrificio*. In this work, Proclus outlines the principles of sympathetic magic, noting that (in Ficino's words) 'from every single star...

---

[15] See for example Edgar Wind, Botticelli's Primavera', *Pagan Mysteries*, pp. 113–27.

[16] See Marsilio Ficino, *Three Books on Life*, eds. Carol V. Kaske and John R. Clark (Binghamton, NY: Medieval & Renaissance Texts & Studies, 1989), p. 263.

[17] Ficino, *Three Books*, III.vii, p. 275.

[18] See Ficino, *Three Books*, vi, p. 265.

[19] See Wind, *Pagan Mysteries*, pp. 36–52.

there hangs its own series of things down to the lowest'.[20] Gathering together several substances belonging to a particular planet creates an item of power, which can be used either to attract or repel.[21] Ficino also incorporates elements of the statue magic described in the *Corpus Hermeticum, Chaldean Oracles* and Iamblichus's *De Mysteriis*, in which a divine being is lured into an image.[22] In addition, he drew upon a Latin translation of the Arabic magical text *Picatrix*, which includes detailed instructions for engraving talismans with sigils of the constellations and the mansions of the Moon.[23]

Ficino gives an example from his own experience of making a talisman to attract the influence of the stars of the Great Bear, which are considered powerfully magical because in moderate Northern latitudes they are always visible. The young Ficino engraved an image of the constellation on a lodestone and hung it from an iron chain, but he found 'that the influence of that constellation is very Saturnine and Martial. I learned from the Platonists that evil demons are mostly Northern…' Saturn and Mars are of course the two malefics whose energies he was keen to avoid.[24]

However, making talismans to attract, for example, solar power is, Ficino assures his readers, perfectly safe. He recommends making a pendant from solar precious stones such as garnet and chrysolite, set in gold and hung on threads of yellow silk. The pendant should be made 'when the Sun is in Aries or Leo and is ascending or else occupies the mid-heaven and aspects the Moon'.[25] On it should be engraved 'a king on a throne in a yellow garment and a raven and the form of the Sun'.[26] This talisman is suitable for those who are under threat from Saturn or Mars, whether natally or by transit. A lunar talisman can be made of selenite set in silver and suspended from a silver thread, created when the Moon is entering Taurus or Cancer and either ascending or at the midheaven.[27]

---

[20] Ficino, *Three Books*, III.xiv, p. 309.

[21] See Proclus, 'On the Sacred Art' in *De Mysteriis Aegyptiorum*, ed. S. Ronan (Hastings: Chthonios Press, 1989), pp. 146–49.

[22] See G. R. S. Mead, *Thrice Greatest Hermes: Studies in Hellenistic Theosophy and Gnosis* (Charleston, SC: Nabu Press, 2010), 2.221; Ruth Majercik, *The Chaldean Oracles: Text, Translation and Commentary* (Leiden: Brill, 1989) fr. 224, p.137 and Iamblichus, *De Mysteriis*, trans. Emma C. Clarke, John M. Dillon and Jackson P. Hershbell (Atlanta: Society of Biblical Literature, 2003), III.28, pp.189–93.

[23] See *Picatrix: The Latin Version of the* Ghayat al-hakim, ed. David Pingree (London: Warburg Institute, 1986), pp.14–21.

[24] Ficino, *Three Books*, III.xv, p. 317.

[25] Ficino, *Three Books*, III.xv, p. 315.

[26] Ficino, *Three Books*, III.xviii, p. 337.

[27] See Ficino, *Three Books*, p. 315.

The most powerful image, Ficino says, is the cross. The Egyptians used it in the form of the ankh long before Christians adopted it, but it derives its fundamental power from the cosmos, because planets on the four angles of a horoscope are particularly strong, and the cross thus most fully absorbs 'the forces and spirits of the planets'.[28] Jewellery engraved or decorated with a cross could of course pass for Christian while channelling cosmic harmony to the wearer.

Following Proclus's principle of gathering as many related items as possible, the use of talismans should be supplemented with appropriate planetary foods, incense, flowers and herbs. The right colours should be worn and the company of children of the planet sought out. Planetary activities should be pursued and the appropriate subjects made matter for thought. In these ways, the Ficinian magician becomes increasingly aligned to the divine force represented by the planet in question. For example, the benevolent power of the Sun is present in:

> flowers which are called heliotrope because they turn towards the Sun, likewise gold,…golden colours, chrysolite, carbuncle, myrrh, frankincense, musk, amber,… yellow honey…cinnamon…; the ram, the hawk, the cock, the swan, the lion… and people who are blond, curly-haired, prone to baldness and magnanimous.[29]

Ficino's whole system of harmonising the soul can be visualised in the image of a cosmic diagram. Although the Sun, Jupiter and Venus are Ficino's favourite sources of power, the human microcosm contains all seven planets.[30] Ficino's principle of harmony means that life is a delicate matter of keeping the varying influences balanced: hot Sun balances cool Moon; Venus balances Mars; Jupiter, Saturn; while Mercury is neutral.

Ficino sets out a system of rulerships that encompasses every area of life. It can be seen in a double focus: from one viewpoint, each planet needs some level of attention every day; but they also represent a seven-runged ladder leading step by step from care of the body to spiritual union with God. It is significant that this astrological ladder begins and ends with the use of images, confirming Ficino's profound involvement with aesthetics.

The Moon rules stones and metals and the images engraved on them, and hence talismans.[31] To Mercury belong the ingredients of natural medicine such as fruits, plant extracts, herbs and gums. Since Ficino sees food and drink as medicinal, these

---

[28] Ficino, *Three Books*, III.xviii, p. 335.

[29] Ficino, *Three Books*, III.i, p. 249.

[30] See for example Ficino's classic account of the inner planets in *Letters*, vol.4 (1988), pp. 61–63.

[31] Ficino, *Three Books*, III.xxi, pp. 355, 357.

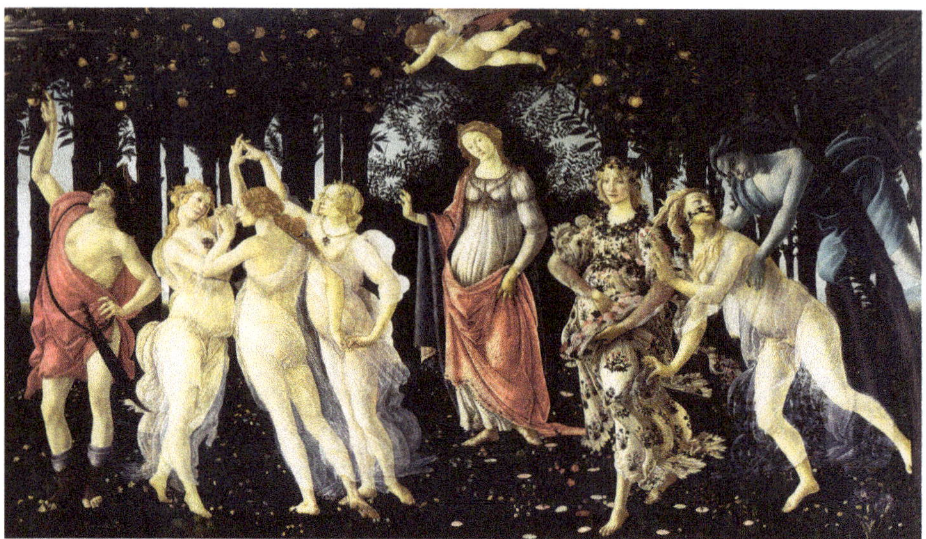

Figure 6.3: Botticelli, *Primavera*, c.1482, Uffizi Gallery.

too come under Mercury, and they are to be used to balance the four humours in the body. Mercury is also the god of astrology and magic.[32]

Venus rejoices in the odours of flowers and incense. Since scent is invisible, Ficino thinks of it as a bridge between body and spirit. He makes a variety of suggestions for using scents: cinnamon, aniseed and fennel warm and cheer the spirit, whilst fresh toast, mint or citron all aid alertness. Odours of decay should be avoided and fresh country air sought out.[33] And of course, the correct incense should be used when invoking the planetary deities.

Holding the middle place in the Chaldean order of planets is the Sun, whose god Apollo gives us music. Ficino includes under this heading 'gestures of the body, dancing, and ritual movements…'[34] and I would argue that paintings with figures dancing or gesturing may well be designed to attract solar influence.[35]

For example, Botticelli's *Primavera* includes figures of the Graces dancing, and Venus making a gesture of welcome to the painting's viewer (FIGURE 6.3). These figures have the additional virtue of benefic planetary energy, the Three Graces representing the three beneficent planets and Venus, of course, herself. Botticelli's paintings have, of course, long been linked by art historians such as Ernst Gombrich and

---

[32] Ficino, *Three Books*, III.xxi, p. 357.

[33] Ficino, *Three Books*, III.xi, pp. 291, 293.

[34] Ficino, *Three Books*, III.xxii, p. 363.

[35] On gesture in paintings, see Michael Baxendall, *Painting and Experience in Fifteenth-Century Italy* (Oxford: Oxford University Press, 1972), pp. 60–71.

Edgar Wind to Ficino's letters and writings.[36] Botticelli worked on some projects with the Pollaiuolio brothers and like Ficino was patronised by the Medici, so he and Ficino certainly crossed paths. There is not room here to digress into the lengthy history of interpretation of the painting, which has both veered away from and back towards a Neoplatonic reading, nor to enter into discussion of the complexities of Medici patronage of Ficino. Suffice it to say that although Lorenzo de' Medici distanced himself from Ficino following the Pazzi conspiracy in 1478, Ficinian ideas remained influential in the Medici circle throughout the period when Botticelli created his mythological paintings.[37]

The next rung on the planetary ladder is that of Mars, which slightly surprisingly represents the first step into pure mind. Mars rules 'strong concepts of the imagination', the mental passions from which action springs.[38] The creative imagination of the artist needs Martial enthusiasm.

Jupiter is the next sphere, ruling the higher realm of mind, 'the sequential arguments and deliberations of the human reason…'[39] But Jupiter also rules the hearty enjoyment of good food, drink and company, and we know that Ficino's preferred milieu for lively debate was a symposium. Plato's *Symposium* evokes the erotic and inebriated atmosphere of such banquets, and Plato himself was alleged to have died after a dinner party held to celebrate his 81st birthday, suggesting a very wild time indeed. Here there is an aesthetic connection in the appreciation of the beauty of young men, a predilection that Ficino shared with Plato's master Socrates. Although those who are uninitiated into the techniques of Platonism might be surprised at this apparently carnal intrusion into the spiritual ladder to divinity, those who have read and understood the *Symposium* will recognise that the path of love finds its inspirational beginning in the divine glow of beauty emanating from some human bodies. Ficino describes the process succinctly in one of his letters. 'The soul, consumed by the divine brilliance which shines in the beauteous man as though in a mirror, is seized unknowingly by that brilliance, and is drawn upwards as by a hook, so that the soul becomes God'.[40]

The seventh and final rung belongs to Saturn, which orbits nearest to the sphere of fixed stars and hence represents the approach to God. Here all material concerns

---

[36] See E. H. Gombrich, 'Botticelli's Mythologies: a study in the Neoplatonic Symbolism of his circle', *Symbolic Images: Studies in the Art of the Renaissance* (London: Phaidon Press, 1972), pp. 31–81 and Wind, *Pagan Mysteries*, pp. 113–27.

[37] See Ames-Lewis, 'Neoplatonism', pp. 328–31.

[38] Ficino, *Three Books*, III.xxi, p. 357.

[39] Ficino, *Three Books*, III.xxi, p. 357.

[40] Ficino, *Letters*, vol.1 (1975), p. 85.

are left behind to focus on the spiritual world in absorbed meditation, so Saturn rules 'tranquil contemplations of the divine.'[41] And here Ficino returns to images.

Ficino suggests focusing one's meditation on an image of the cosmos, 'an archetypal form of the whole world…'[42] He seems to have in mind a working model of the solar system and the fixed stars, inlaid with metals and precious stones such as gold and lapis lazuli. Ficino instructs his readers that the Earth and Venus should be green, Jupiter blue, and the Sun gold. These three colours should also be used for the rest of the model, because they will imbue it with all the virtues of the Three Graces. The model should be made either on Sundays or during planetary hours of the Sun, and finished at the moment when the Sun first enters Aries. However, if it proves impossible to make such an elaborate and expensive item, a private room should be frescoed with the cosmic image in the Graces' colours.

The meditator either retires to the room to contemplate the fresco or model, or carries the model around. Ficino recommends that the meditation be performed as often as possible, and one should sleep in the frescoed room so as to continue absorbing its magical influence. The image serves to increase awareness of the cosmos within, so that the inner planetary orbits are brought into harmony with the outer ones. Thus becoming 'very orderly and temperate in…thoughts, emotions, and mode of life,'[43] the meditator purifies and perfects his or her soul.

Ficino's cosmic image represents talismanic art taken to its highest form, in a Proclian sense drawing to the meditator not just one influence but all the powers of the cosmos in a divine harmony. To conclude, let us consider two paintings that suggest a talismanic purpose.

Botticelli's *Birth of Venus, Venus and Mars* and the *Primavera* are traditionally interpreted as allegorical reflections on beauty and peace, but I suggest that given Ficino's emphasis on the spiritual value of beauty and his magical practices these are talismanic paintings to attract the power of the benefic planet Venus. In *Venus and Mars* the astrological link is absolutely clear (FIGURE 6.4).

Venus and Mars has been described as illustrating Ficino's claim in his commentary on Plato's *Symposium* that 'Venus often shackles…the malignancy of Mars, by coming into conjunction or opposition with him, or by receiving him or by watching him from the trine or sextile aspect….*But Mars never dominates Venus.*'[44] The Italian

---

[41] Ficino, *Three Books*, III.xxi, p. 357 and xxii, p. 363.

[42] Ficino, *Three Books*, III.xix, p. 345.

[43] Ficino, *Three Books*, III.xix, p. 347.

[44] Marsilio Ficino, *Commentary on Plato's* Symposium *on Love*, trans. Sears Jayne, (Woodstock, CT: Spring, 1985), p. 97 (translator's emphasis).

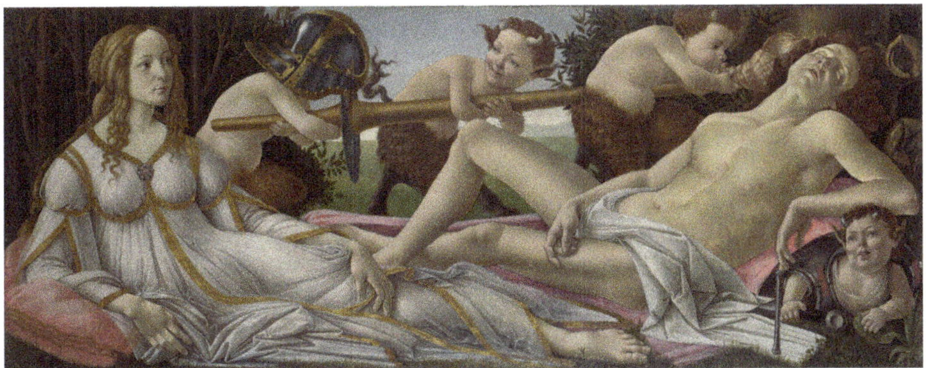

Figure 6.4: Botticelli, *Venus and Mars*, c.1483, National Gallery.

version of D*e amore* was circulating in manuscript by 1474, so Ficino's work enjoyed wide distribution even amongst those who had no Latin and it was mostly received with enthusiastic praise amongst his friends and admirers.[45]

The painting appears to illustrate a separating conjunction, sexual union having taken place. But it can also be said that Venus has received Mars (and dominated him sexually enough to wear him out). Also, the angle between the upper side of the two bodies is 120° – a trine – from which aspect Venus is very coolly watching Mars. Close inspection of the painting can also tease out sextile angles and the opposition; the apparently simple design in fact appears to conceal the kind of hidden complexities that so delighted aristocratic patrons of art at this period.

Ficino's astrological statement about the greater strength of Venus is based on observed astronomical fact. Whenever both planets are visible in the night sky Venus, being the most brilliant planet as seen from Earth, will always far outshine Mars. So Botticelli's painting presents a beautiful and amusing image to illustrate an astronomical fact that is interpreted astrologically, thereby confirming the superior power of love, harmony and peace over war and disharmony.

As a contrast, I suggest Antonio Pollaiuolo's *Battle of the Nude Men* as a talisman against the malefic principle of Mars let loose (FIGURE 6.5). This circular battle of men with their faces distorted by hatred is designed to show that none of them will survive; the battle will end in total annihilation. In other words, the martial energy destroys itself. I propose that the engraving illustrates the use of talismanic art to repel a destructive, disharmonious force; it is apotropaic.

Beauty, for Ficino, was the gateway to the Divine, and the creative powers of

---

[45] See Ficino, *Commentary*, p. 20.

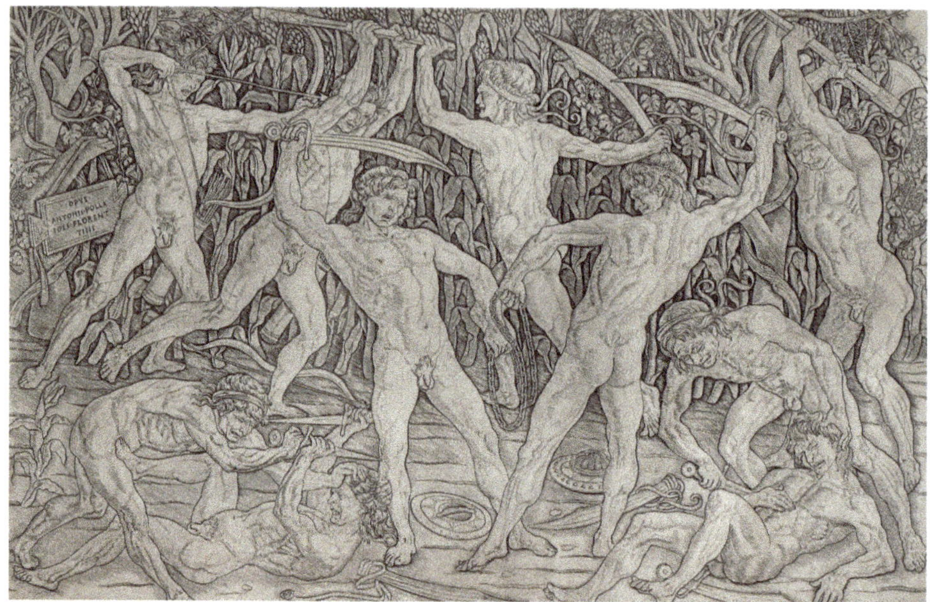

FIGURE 6.5: Antonio Pollaiuolo, *Battle of the Nude Men*, c.1470s, British Museum.

man expressed in art, architecture and music brought him nearer to God because through exercising them he imitated the nature of his Creator. Ficino's involvement with astrology and magic mingled inextricably with his Christian faith, itself profoundly informed by the magical religious rites of the Neoplatonists. The continuing power of Botticelli's images of Venus may well be due as much to Ficino's intimate knowledge of such esoteric sources as to the artist's exquisite technical expertise.

Lower Astrology and Silent Poetry:
John Donne's Portrait

Spike Bucklow

ABSTRACT: This chapter focuses on a rather exclusive late-sixteenth-century in-joke. As such, the cultural impact of the chapter's subject was probably quite minimal. However, its existence provides evidence of a sophisticated knowledge of astrology as an everyday frame of reference available to be harnessed in play and with a reasonable expectation of recognition. The chapter considers (in the etymologically strict sense) the use of an everyday painting material in a circa 1595 portrait of John Donne. The metaphysical poet commissioned this iconic work (now in the National Portrait Gallery) from an unknown painter as a carefree young man about town. In later life, he had it in his possession as a widower and the widely respected Dean of St Paul's Cathedral. The aspect of the painting I wish to examine is a meditation on love-sickness.

John Donne, the metaphysical poet, commissioned a number of portraits. This chapter considers the one he described as being 'in shadows' (FIGURE 7.1).[1] The portrait was painted around 1595 and it features absolutely nothing explicitly related to astrology. Whilst the portrait is indeed dark and shadowy, alternate interpretations of the word 'shadow' provide clues to suggest how the portrait's presence might be justified in a volume dedicated to astrology.

The word 'umber' means shadow and, in two plays of around 1599, Donne's contemporary, William Shakespeare, used that word to denote both material and immaterial phenomena. In *As You Like It*, Celia said she would disguise herself '...in poor and mean attire / And with a kind of umber smirch my face'.[2] But, in *Henry V*, troops had 'umbered faces' on the eve of battle not because they were smeared with the earth of Umbria (or even the earth of Agincourt), but because they were shadowed by flickering pre-dawn camp-fires.[3] Around the turn of the seventeenth century, umber meant both an immaterial shadow and a material with which shadow-like effects could be produced. In addition, Shakespeare knew that playactors themselves could be considered as 'shadows', in the sense implied by Plato's myth of the cave.[4] Indeed the actors refer to themselves as 'shadows' and, since 'All the world's a stage', every one of us could be considered a shadow.[5] Since Plato implied that his cave dwellers

---

[1] R. C. Bald, *John Donne, A Life* (Oxford: Clarendon 1986), p. 567.

[2] *As You Like It* (I, iii, 107–8).

[3] *Henry V* (IV, prologue, 9).

[4] Plato, *Republic*, (VII, 514a–518e) tr. P. Shorey (London: Heinemann 1963), 2, pp. 119–37. *Oxford English Dictionary*, 'shadow, n', online version, accessed 27 March, 2018.

[5] *A Midsummer Night's Dream* (Epilogue, 1) and (V, i, 414–19); *As You Like It* (II, vii, 138).

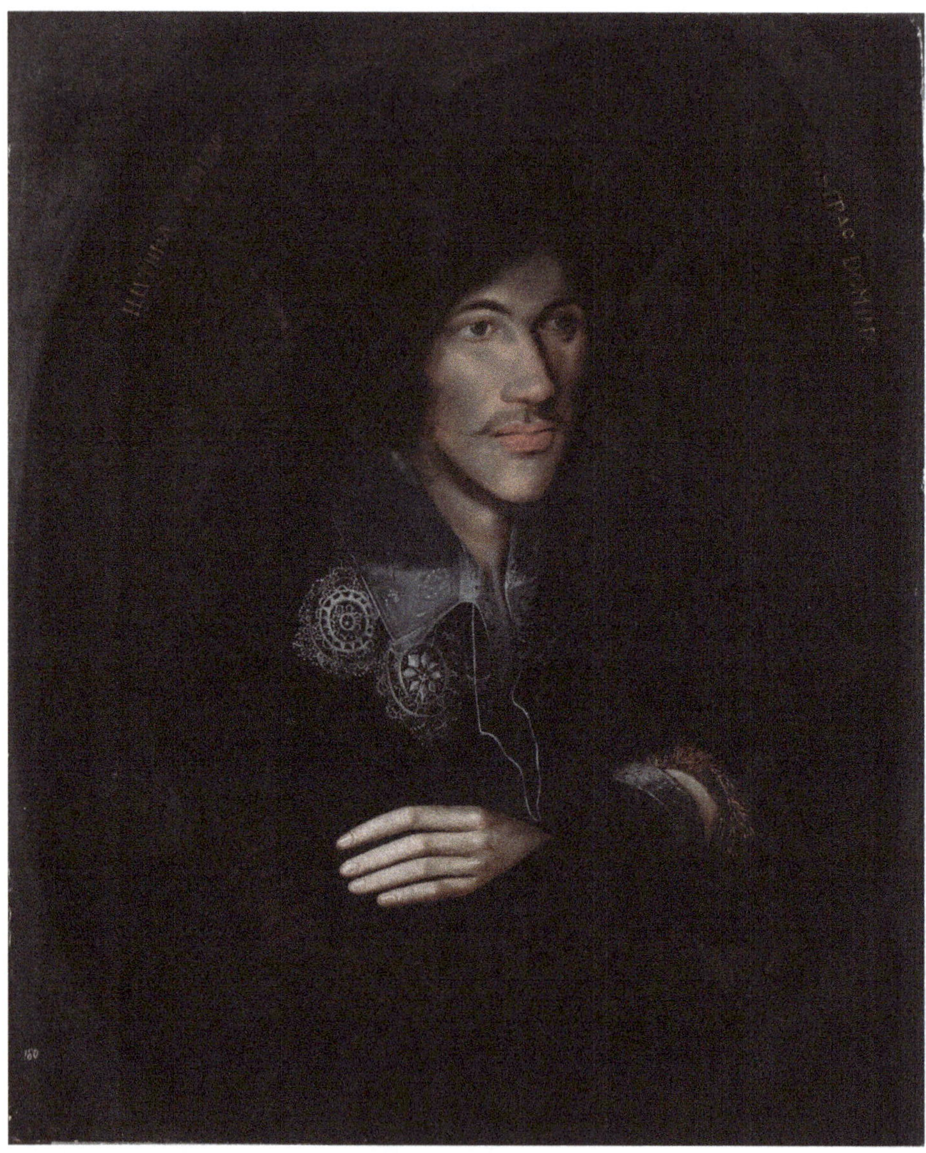

FIGURE 7.1: John Donne, by an unknown English artist oil on panel, circa 1595 copyright National Portrait Gallery, London (NPG 6790) Chris Titmus / Hamilton Kerr Institute, University of Cambridge.

lived as if in a dream, shadows were also synonymous with dreams, and therefore again with all of us, since 'We are such stuff as dreams are made on'.[6]

Donne was a 'great frequenter of plays' who may have known Shakespeare personally (both frequented the Mermaid and they had a good mutual friend in Ben Jonson).[7] Donne also confessed to an 'immoderate desire of humane learning' so would have been very familiar with the multiple meanings of the word 'shadow' as applied to portraiture and elaborated upon in contemporary poetry and literature.[8] In everyday Elizabethan use, 'shadow' could mean the relative darkness cast by light, a copy, a representation, a portrait, an actor, a vain pursuit, a ghost, a pre-figuration and more.[9] In his mournful poem about a man giving his portrait to his lover upon their parting, Donne referred to a time 'when we are shadows both', and in another poem he noted the inscrutability of 'the shadow' worn by women.[10] He also played with details of the optical phenomenon in *A Lecture Upon The Shadow*.[11]

Everyday word-play on 'shadow' indicates a fluid relationship in late-sixteenth-century English between the material and immaterial, and the animate and inanimate. This is a direct consequence of the pre-modern worldview, for it has been said that '[i]t was easy … for the middling-thoughtful man to be Platonist; in fact it was difficult for him not to be'.[12] In the Neo-Platonic sense, a shadow is any projection or reflection of an Ideal. As such, the painted portrait and the living sitter may both be considered merely partial manifestations of the sitter's essence, one 'shadow' being relatively immutable whilst the other was all too mortal.

One of the materials with which Donne's portrait was painted was an earth pigment called 'umber', the material Celia used to disguise herself. Appropriately, it was used in the portrait's dark, shadowy background. However, the portrait's astrological connection emerges from the identity of another pigment. It is to be approached via

---

[6] *Tempest* (IV, i, 156–57) This useage survives, for example, 'That eternal world, whereof all here is but a shadow and a dream', C. Kingsley, *Westward Ho* (Leipzig: Bernhard Tauchnitz, 1855), vol. II, p. 53.

[7] Richard Baker, *Chronicles*, (1643), 156, as quoted in J. Stubbs, *Donne, the reformed soul* (London: Penguin 2007), p. 47.

[8] *Letters to Severall Persons of Honour* (1651), p. 51. as quoted in Stubbs, *Donne*, 4; C. Pace, 'Delineated lives; themes and variations in seventeenth-century poems about portraits', *Word and Image* 2, no. 1 (1986): p. 4.

[9] A. Gash, 'Shakespeare's comedies of shadow and substance; word and image in *Henry IV* and *Twelfth Night*', *Word and Image* 4, nos. 3–4, (1988): pp. 626–62.

[10] *Elegy V, His Picture* in John Donne, *Complete Poems*, ed. H. Faussett, (London: Dent 1974), p. 63; *Twickenham Garden*, in Donne, *Poems*, p. 17.

[11] Donne, *Poems*, p. 50.

[12] R. L. Colie, *My Echoing Song: Andrew Marvell's Poetry of Criticism* (Princeton: Princeton University Press, 1970), p. 298.

the discipline of alchemy, a discipline that Donne drew upon in his love poetry.[13] His use of alchemy involved quite detailed analogies, for example,

> 'Then like the Chemic's masculine equal fire,
> Which in the Limbec's warm womb doth inspire
> Into th' earth's worthless dirt a soul of gold'[14]

Lower astrology

Alchemy was sometimes called 'lower astrology', drawing upon the fact that much alchemy involved the manipulation of metals and in recognition of the fact that metals were traditionally connected with planets.[15] This was such a commonplace and even artists' recipes referred to metals by their planetary equivalents.[16] Donne would have known Chaucer's *Canterbury Tales* which contains one of many popular expressions of the correlation between metals and planets.

> Gold for the Sun and Silver for the Moon,
> Iron for Mars and Quicksilver in tune
> With Mercury, Lead which prefigures Saturn,
> And Tin for Jupiter. Copper takes the pattern
> Of Venus if you please.[17]

In Elizabethan terms, metals were the 'shadows' of planets. Shadows – and those other optical projections, reflections – transmit some aspects of an entity, and that which is transmitted necessarily has a more limited and contingent reality than the entities from which they are projected. For example, we have solid bodies but our shadows are flat, and our bodies are relatively durable whilst our shadows are modified by the terrain upon which they fall and they disappear completely in the dark. Shadows and reflections loose some aspects (solidity), they project some aspects (like silhouettes) but, crucially, they also invert some aspects (left and right) of their counterpart further up the Great Chain of Being.[18]

---

[13] *Love's Alchemy*, in Donne, *Poems*, p. 25.

[14] *Elegy VII, The comparison*, in Donne, *Poems, p.* 66.

[15] M. P. Merrifield, *Original Treatise on the Arts of Painting*, (New York: Dover, 1967), I, pp. 66–68.

[16] See for example, A. Roob, *The Hermetic Museum, Alchemy and Mysticism* (Cologne: Taschen, 1997), pp. 80, 665, 673.

[17] Chaucer, *The Yeoman's Tale*, II.

[18] A. O. Lovejoy, *The Great Chain of Being* (Harvard: Harvard University Press, 1976).

Whilst the planets were 'intelligent bodies', the metals were 'embodied intelligences'.[19] As celestial bodies, the planets were in various states of 'Being' whilst, as terrestrial embodiments, the metals were in various states of 'becoming'. Those planets and metals that were connected had analogous ways of participating in their respective – celestial and terrestrial – levels of reality. So, of the seven traditional planets, one was luminous (the Sun) whilst the other six were reflective. Likewise, of the seven traditional metals, one was incorruptible (gold) whilst the other six were corruptible. The six corruptible metals were seen as various stages in the process of 'becoming' gold and it follows that, of all terrestrial materials, incorruptible gold participated most in the celestial property of 'Being', thus transcending the other metals as the Sun transcended the other planets.

This chapter considers one of the corruptible metals (lead, connected with Saturn) and, in particular, looks at the product of the metal's corruption – its rust. The corruption of a metal can be seen as a further step down the Great Chain of Being toward ever-more relativity and contingency. The consequence of metal's corruption, rust, is another 'shadow' that transmitted some aspects whilst inverting others. For example, iron, the shadow of Mars, was (appropriately enough) used to make martial weapons that drew blood. However, rust, the shadow of iron, was used to stem the flow of blood. Both the metal and its rust had relationships with blood yet one caused bloodshed whilst the other cured bloodshed.[20]

In the traditional Ptolemaic-based systems, Saturn was the highest planet and, following the law of inversion, it was connected to the lowest, or basest, of the corruptible metals, lead. (Likewise, the Moon was the lowest planet and it was connected to the highest, or noblest, of the corruptible metals, silver.) Again following the law of inversion, lead was a black metal and when corrupted it produced a white rust. (Likewise, silver was a white metal and when corrupted it produced a black rust.) As well as residing at a lower level of reality, the rust of a metal can also be thought of as showing its inner nature or its 'true colour'. The true colour of outwardly black lead would therefore be white, the purest of all colours, and this relationship is also expressed in the alchemical adage 'the inner nature of lead is gold' (just as 'the inner nature of gold is lead').[21]

As individuals, we tend to show our 'true colours' when tried by adversity. Similarly, the coloured rusts were produced by subjecting metals to trials. The traditional understanding of metals assumed they were composed of, amongst other elements,

---

[19] T. Burckhardt, *Alchemy* (London: Stuart and Watkins, 1967), p. 77.
[20] S. Bucklow, *The Alchemy of Paint* (London: Marion Boyars, 2009), pp. 109–40.
[21] Bucklow, *Alchemy*, p. 273.

water, and their malleability (which reflected the planets' movements, in contrast to the stone-like fixed stars) was one expression of their watery nature. Powdery rusts were therefore formed by drying-out metals with fire. Incorruptible gold was the only metal to survive trial by fire, and each of the other metals rusted when exposed to fire in various forms. Silver, for example, tarnished when exposed to sulphur or 'brimstone', a mineral embodiment of fire related to the legendary Hell-fire. Of course, silver's corruption occurred accidentally and could be remedied by polishing, as befitted a relatively 'noble' metal. This apparently reversible cycling between reflective and tarnished states was a projection of the Moon's phases which could also be seen in the bright and occluded, or the manic and depressive, phases of 'lunatic', or bipolar, individuals.[22]

Lead was the basest of metals and its irreversible rusting occurred when it was exposed to the heat of decay and putrefaction.[23] This trial could occur accidentally, and one seventeenth-century artists' manual recommended scraping white rust off lead roofing that lay directly on oak beams in old buildings (oak was a source of 'hot' acidic vapours).[24] However, the rusting of lead was also undertaken on purpose in order to use the white powder as a pigment in paintings. A fifteenth-century artists' manual described the white pigment simply as being 'made by alchemy'.[25] A thirteenth-century artists' manual described the manufacture of this same white pigment in detail – coils of lead metal were placed under horse dung for a month after sprinkling them with vinegar and urine.[26] These technicalities would have been known to whoever painted the portrait and they would also have been known to an educated Elizabethan patron like John Donne.

John Donne

John Donne was born in 1572, the son of an ironmonger, and his very early life would have been surrounded by metals. When he was four years old his father died and his mother married a physician, a profession that was at the time closely allied to alchemy. He matriculated at Oxford in 1584, may then have moved on to Cambridge

---

[22] S. Bucklow, *The Riddle of the Image* (London: Reaktion, 2014), pp. 193–95.

[23] Bucklow, *Riddle*, pp. 11–41.

[24] John Smith, *The Art of Painting* (1687), 16. as quoted in I. Bristow, *Interior House Painting Colours and Technology* (New Haven: Yale University Press, 1996), p. 10.

[25] Cennino Cennini, *The Craftsman's Handbook* (59), trans. D. V. Thompson (New York: Dover, 1960), p. 34.

[26] Theophilus, *On Divers Arts* (III, 37), trans. J. G. Hawthorne and C. S. Smith (New York: Dover, 1979), pp. 41–42.

and was at the Inns of Court by 1592. The following year he received, and spent most of, his inheritance and his brother died in prison. In 1596 he sailed with the Earl of Essex and triumphantly attacked Cadiz harbour. The following year he joined another expedition with Sir Walter Raleigh, but returned unsuccessful. Back home, he gained employment with Sir Thomas Egerton, lord keeper of the great seal.[27]

This brief summary of the early part of Donne's life hides a profound change – he was born into a Catholic family (which accounts for his very early matriculation and his brother's death). But by the age of 24 he was sailing against Catholic Spain and was shortly to enter the Church of England (employment by Egerton and advancement in the establishment would have been very difficult as an open Catholic). Donne's poetry is difficult to date, but a significant portion of it was probably written in the 1590s.

The portrait 'in shadows' was probably painted around 1595, whilst Donne was in his early- to mid-twenties and briefly entertaining a military career. It was his second portrait. The first portrait was painted in 1591, when he was nineteen years old, and is known only through an engraving in the 1635 edition of Donne's *Poems*.[28] Its painter is unidentified but it has been suggested that the original may have been by Nicholas Hilliard, mainly on the strength of Donne's reference to Hilliard in *The Storme* (probably written upon return from his second expedition to Spain). The poet's third portrait was painted by Isaac Oliver in 1616. It too was engraved and it appeared in the 1640 edition of Donne's *LXXX Sermons*.[29] A fourth, of 1620, by another unknown painter, is in a very unusual format that relates to classical medallion portraits. It was engraved and was published in the 1651 edition of Donne's *Letters*.[30] The last, lost, portrait was painted in 1631 but its uppermost portion is known through an engraving published in *Death's Duel* of 1632.[31] The whole painting served as an approximate model for Nicholas Stone's 1632 funeral monument for Donne in St Paul's Cathedral.

It has been rightly said that, with the exception of the Oliver miniature, all these images are 'out of line with the ordinary portraits of the time, and have a marked element of role-playing'.[32] (Of course, in Elizabethan terms, such role-playing identifies the sitter as an actor or 'shadow' in most of his portraits, not just the one

---

[27] D. Colclough, 'John Donne' at http://www.oxforddnb.com [Last accessed 18 September 2018].

[28] D. Piper, *The Image of the Poet: British Poets and their Portraits* (Oxford: Clarendon 1982), p. 25.

[29] Piper, *Image*, pl. 21, p. 27.

[30] Piper, *Image*, pl. 24, p. 29.

[31] Piper, *Image*, pl. 28, p. 31.

[32] Piper, *Image*, p. 28.

considered in this chapter.) This role-playing is perhaps most evident in the final portrait, the genesis of which was reported by Donne's biographer, Izaak Walton. According to him, Donne rose naked from his death bed, wrapped himself in a loose shroud – which was knotted at his head, allowing his face to show – and stood on an urn facing east 'from whence he expected the second coming of his and our Saviour Jesus'. The shroud in this painting would have looked like a flame issuing from the urn. His face at the top of the shroud-flame would have made him appear to rise phoenix-like from his ashes.[33] For this picture, he chose an un-named 'choice Painter' who drew him on a full-length board. Donne kept this portrait in his bedroom, where it became his 'hourly object till his death' about three weeks later on 31st March, 1631.[34]

Donne evidently felt that painted images were sufficiently important to employ at least one, and possible two, of the best portrait painters of the age – Isaac Oliver and maybe also Nicholas Hilliard. However, he also employed completely unknown painters for his shroud painting and for the painting 'in shadows'. This may be significant since choosing a lesser painter could have allowed him to exert a greater influence over the painting process. In wishing to control the painting process, Donne would have been following the example of a close friend and fellow (but neglected) metaphysical poet, Lord Herbert of Cherbury.[35] Herbert commissioned several portraits that appear to have been composed to his own programme, in the first coordinated attempt by a poet to manipulate his painted image.[36]

Donne may not have held the paint brush but he should be seen as the painting's 'creative director' and the portrait's intellectual content undoubtedly came from the poet. Today, painting and poetry seem to be very different art-forms yet, in the late sixteenth century, they had much in common.

Silent poetry

English Renaissance portraiture was deeply informed by the simile *ut picture poesis*, 'as is poetry, so is painting', which seventeenth-century art theorists could also read 'as is painting, so is poetry'.[37] Poetry and painting were sister arts and

---

[33] See, however, *First Anniversary*, pp. 215–16.

[34] Izaak Walton, *Lives of Dr John Donne, Sir Henry Wooton, Mr Richard Hooker, Mr. George Herbert* (Menston: Scholar Press facsimile, 1969), p. 75.

[35] R. D. Bedford, *The Defence of Truth, Herbert of Cherbury and the Seventeenth Century* (Manchester: Manchester University Press, 1979), pp. 10–11.

[36] Piper, *Image*, p. 24.

[37] Horace, *Ars poetica*, p. 361, as quoted in R. W. Lee, 'Ut pictura poesis; the humanistic theory of

Lomazzo, in an artist's manual translated into English in 1595, even claimed they were twins.[38] Pictures were silent poems and poems were spoken pictures. Metaphysical poets were amongst those who purposefully blurred the boundaries between the two art-forms.[39] However, pictures were generally considered inferior to poems and, in the hierarchy of visual arts, portraiture was inferior to history painting, as is clear in Donne's lines; 'a hand or eye By Hilliard drawn, is worth an history, by a worse painter made'.[40] Since the most valued part of any painting was its poetic aspect, it is worth considering the relationship between Donne's portrait and his poetry. Whilst the young Donne devoted himself to 'spoken painting', his forays into commissioning and stage-managing 'silent poetry' indicate an interest in the visual arts that was reinforced by his statement later in life that sight was the 'noblest of the senses'.[41] Many of his poems and letters contain references to the portrayal of faces, his own and other's, whether engraved on the heart, reflected in an eye, or conventionally painted.[42]

In a milieu in which words and images were compared more than contrasted, Elizabethan paintings often contained carefully-chosen inscriptions. This portrait is no exception and contains the inscription 'ILLUMINA TENEB[RAS] NOSTRA DOMINA', or 'Lighten my darkness, our Lady'. The portrait also contains visual effects that parallel some of the verbal effects – like puns – employed in his poetry. For example, the inner garment, of which only one sleeve is visible, is dull-brown in colour. In contemporary terminology, this sleeve was 'dunne' coloured. (This colour-term, via 'dynne', accounts for the modern adjective 'dingy', so the whole dingy painting alludes to the sitter's name.[43]) Also, his shirt is undone, reflecting the state of a person 'undone' by separation from their loved one. This particular painted detail visually foreshadows the three-line poem that he allegedly wrote some years later, aged twenty-nine, upon his politically injudicious marriage to his employer's seventeen-year-old daughter.

---

painting', *The Art Bulletin* 22, no. 4 (1940): p. 197; Giovanni Paolo Lomazzo, (VI, 65), 486, as quoted in Lee, 'Poesis', p. 197.

[38] Carles du Fresnoy, *De arte graphica*, as quoted in Lee, 'Poesis', p. 197.

[39] J. Faust, 'Blurring the boundaries; *Ut Pictura Poesis* and Marvell's liminal mower', *Studies in Philology* 104, no. 4 (2007): pp. 526–55.

[40] *The Storm (to Mr Christopher Brooke)*, in Donne, *Poems*, p. 129.

[41] *Easter Day Sermon*, 1628, as quoted in Pace, 'Delineated', p. 4.

[42] *The Dampe*, in Donne, *Poems*, p. 44; *Witchcraft by a picture*, in Donne, *Poems*, p. 30; *His Picture*, in Donne, *Poems*, p. 63.

[43] Oxford English Dictionary, www.oed.com [accessed 21 September, 2015].

John Donne,
Anne Donne,
Undone.'[44]

(This short poem was prescient. John Donne was dismissed from Egerton's service, endured a brief spell in Fleet prison and spent over a decade trying to rebuild his career. There is no evidence that Donne ever regretted following his heart.)

Donne's poetry was littered with puns, so the dingy portrait's dunne sleeve and undone shirt were characteristic of the sitter's written output.[45] Sometimes it was obvious what the poet's words meant and sometimes it was up to the reader or listener to discover what may have been meant.[46] The dunne sleeve and undone shirt would have been relatively obvious to the contemporary viewer but the portrait contains another, more obscure, pun that plays not with the sitter's name but with the melancholic theme that underlies the portrait.

Melancholy

Donne's portrait 'in shadows' was described by a contemporary as being 'in a melancholic posture' and Thomas Morton, who later became Bishop of Durham, reported having seen the picture in 'a dear friend's chamber in Lincoln's Inn all enveloped in a darkish shadow'.[47] Donne was admitted to Lincoln's Inn in 1592, a few years before the portrait was painted and the 'dear friend' to whom Morton referred was probably Christopher Brooke. Donne and Brooke can be linked by their time at Lincoln's Inn and through the poet's verse letter 'To Mr C. B.'. Donne may have been trying to woo Brooke's sister and, if so, the poet's pursuit was apparently unsuccessful since the letter complains;

> 'Yet love's hot fires, which martyr my sad minde,
> Doe send forth scalding sighes, which have the Art
> To melt all ice, but that which walls her heart.'[48]

---

[44] *A Choice Banquet of Witty Jests, Rare Fancies, and Pleasant Novels* (1665), p. 72. as quoted in Stubbs, *Donne*, p. 154.

[45] A. R. Rieke, 'Donne's Riddles', *Journal of English and German Philology* 83, no. 1 (1984): p. 6.

[46] P. A. Jorgensen, 'Much Ado about Nothing', *Shakespeare Quarterly* 5, no. 3 (1954): pp. 287–95.

[47] William Drummond of Hawthornden as quoted in G. Keynes, *A bibliography of Dr John Donne, Dean of Saint Pauls* (Oxford: Clarendon, 1973), p. 373; Richard Badiley, *Life of Morton* (London, 1669), p. 101 as quoted in Keynes, *Bibliography*, p. 373.

[48] D. L. Edwards, *John Donne, man of flesh and spirit* (London: Continuum, 2001), pp. 39–40.

Donne's reference to his second portrait as 'in shadows' is a literal description, in agreement with Thomas Morton's but, as should be expected from his writing, it also has other meanings. It could allude to his melancholy state. Melancholy was a condition of body and soul that was recognisable from an individual's appearance and behaviour. Melancholy was one of the four humours. These were 'sanguine' or hot and wet, which was associated with the element air; 'choleric' or hot and dry, associated with fire; 'phlegmatic' or cold and wet, associated with water; and 'melancholic' or cold and dry, associated with earth. Ideal health involved a balance of all four humours, but imbalances could arise due to lifestyle, diet or natural cycles. Melancholy, for example, was associated with the end of cycles such as autumn, old age or, in Donne's case, with the end of a relationship.

Those dominated by a sanguine temperament were fleshy, ruddy, amiable and courageous whilst choleric individuals were lean, hairy, proud and shrewd and phlegmatic people were short, pale, slothful and dull. Those suffering from melancholy were dry, dark, obstinate and surly with a 'certain slow pace and soft nice gait, holding down their heads, with countenance and look grim and frowning.'[49]

Sorrow was the 'mother and daughter of melancholy, her epitome, symptom and chief cause.'[50] Via the concept of 'hypochondria', melancholy became associated with what is now called depression but, at the time he commissioned his second portrait, John Donne would have seen it as a temporary humoural imbalance or an external quality that could provoke such a temporary imbalance. In other words, melancholy could be a passing mood or something that could induce that mood in everyday life. However, if extreme, a melancholy complexion had the potential to become pathological melancholy that could require treatment with changes in lifestyle, diet or access to psychoactive agents.

The various forms of melancholy were sufficiently well-established to contribute to literature. Chaucer, for example, drew upon the condition in his description of Arcite's melancholy love-sickness.[51] Shakespeare's character Valentine suffered a love-sickness that was as clear to his companions as 'water in a urinal'.[52] By the forth quarter of the sixteenth century, the affectation of melancholy had become

---

[49] Levinus Lemnius, *The Touchstone of Complexions*, trans. Thomas Newton (London, 1576), fol. 146., as quoted in L. Babb, *The Elizabethan Malady* (East Landing: Michigan State College Press, 1951), pp. 9–10.

[50] Burton, *Anatomy* (I, p. 298) as quoted in Babb, *Malady*, p. 24.

[51] *Knight's Tale* (II, 1361–79)

[52] *Two Gentlemen of Verona* (II, i, 35–40).

extremely fashionable, especially among young male aristocrats, recently returned from travels.[53] Donne's allusion to melancholy in his portrait included the clothing in which he chose to be depicted. Apart from the 'dunne' sleeve, the predominant colour is black, which could be the colour of grief and madness (because of the association with melancholy, or 'black bile') although it could also be the colour of constancy.[54] Another allusion was the pose he chose to adopt. In conjunction with a large floppy hat, the downcast face and crossed arms were typical for the sufferer of love-sick melancholy and the portrait's pose is almost identical to that identified as 'Inamorato' on the frontispiece of Robert Burton's *Anatomy of Melancholy*.

Although Donne's melancholy was reflected in his dress and his posture, it was primarily a condition of his soul. Of course, the state of the soul is a relative mystery for most sitters in most portraits, where the appearance of the body is the main focus. However, John Donne was a very self-aware poet, and we are told much about the state of his soul in poetry that is more-or-less contemporary with the portrait.

Donne's soul

The state of the soul was of central concern in metaphysical poetry and it also found expression in Elizabethan portraiture.[55] The soul featured regularly in Donne's poetry as an integral part of his identity, which he could see in terms of the four elements and humours. For example, he described 'My fire of passion, sighs of air, Water of tears, and earthy sad despair', asserting that these were his 'materials'.[56] The portrait in question is of a soul in love, and for Donne, love was 'no quintessence, But mixt of all stuffs'.[57] His masterpiece of love poetry, *The Ecstasy*, contained much imagery of lovers' souls, including the idea of achieving balance through love.

> '... as all several souls contain
> Mixtures of things, they know not what,
> Love these mix'd souls doth mix again
> And makes both one, each this and that.'

---

[53] Babb, *Malady*, p. 76.

[54] Giovanni Paolo Lomazzo, *A Tracte Containing the Artes of Curious Paintinge Carvinge and Building*, (III, 12), trans. R. Haydocke (Farnborough: Greggs, 1970), pp. 113–14.

55 M. Leslie, 'The dialogue between bodies and souls; pictures and poesy in the English Renaissance', *Word and Image* 1, no. 1 (1985): pp. 16–30.

[56] *The Dissolution*, in Donne, *Poems*, p. 44.

[57] *Love's growth*, in Donne, *Poems*, p. 21.

The portrait expresses Donne's hope to be connected to, mixed with, and balanced by, his 'Lady', via his soul – or more accurately, if his love became requited – via their two souls. Lovers' souls were entwined just as, the poem continues, their 'eyebeams twisted', their hands were interwoven, and their souls 'negotiated there' above their prone bodies. The mixing of souls he experienced with his lover was so profound that 'he knew not which soul spake, Because both meant, both spake the same ... soul into the soul may flow'. He described the ecstatic connection between lovers as a

> '... dialogue of one ...
> Our souls (which to advance their state,
> Were gone out) hung 'twixt her, and me.'[58]

For Donne, souls could be a link between bodies and he depicted his soul as just such a linkage in his 'melancholy' portrait in 'shadows'. The painted soul is easily overlooked, but it is the faint wisp of white that emerges from his undone shirt, rises to his throat and disappears into the air. It seems to come from the heart and (to my knowledge) is quite unlike any other portrayal of the soul in the English visual tradition. In illuminated manuscripts, for example, the soul is usually depicted departing the body at death via the mouth and is usually in the form of a small child or homunculus. Donne's 'advancing' melancholy soul gave him the opportunity to engage in the portrait's most obscure pun. This pun drew upon his understanding of souls as mixed and fluid entities.

The pun

The poet's usual materials are words, and puns exploit the polysemic or multivalent nature of words. Words and sounds can therefore potentially carry multiple meanings in 'spoken painting', as demonstrated by the three-line poem that predicted the fall-out from Donne's politically injudicious marriage to his employer's daughter. (Punning pictograms often served similar purposes in heraldic coats of arms.) In the pre-modern world, where connections between things could be spiritual as well as material, meanings could also be ascribed to the inanimate parts of creation and those inanimate entities could be included in very wide webs of meaning. As was said about a decade after Donne's death:

> The finger of God hath set an inscription on all His workes, not graphicall or composed of Letters, but of their severall formes, constitutions, parts and operations,

---

[58] *The Ecstasy*, in Donne, *Poems*, p. 34.

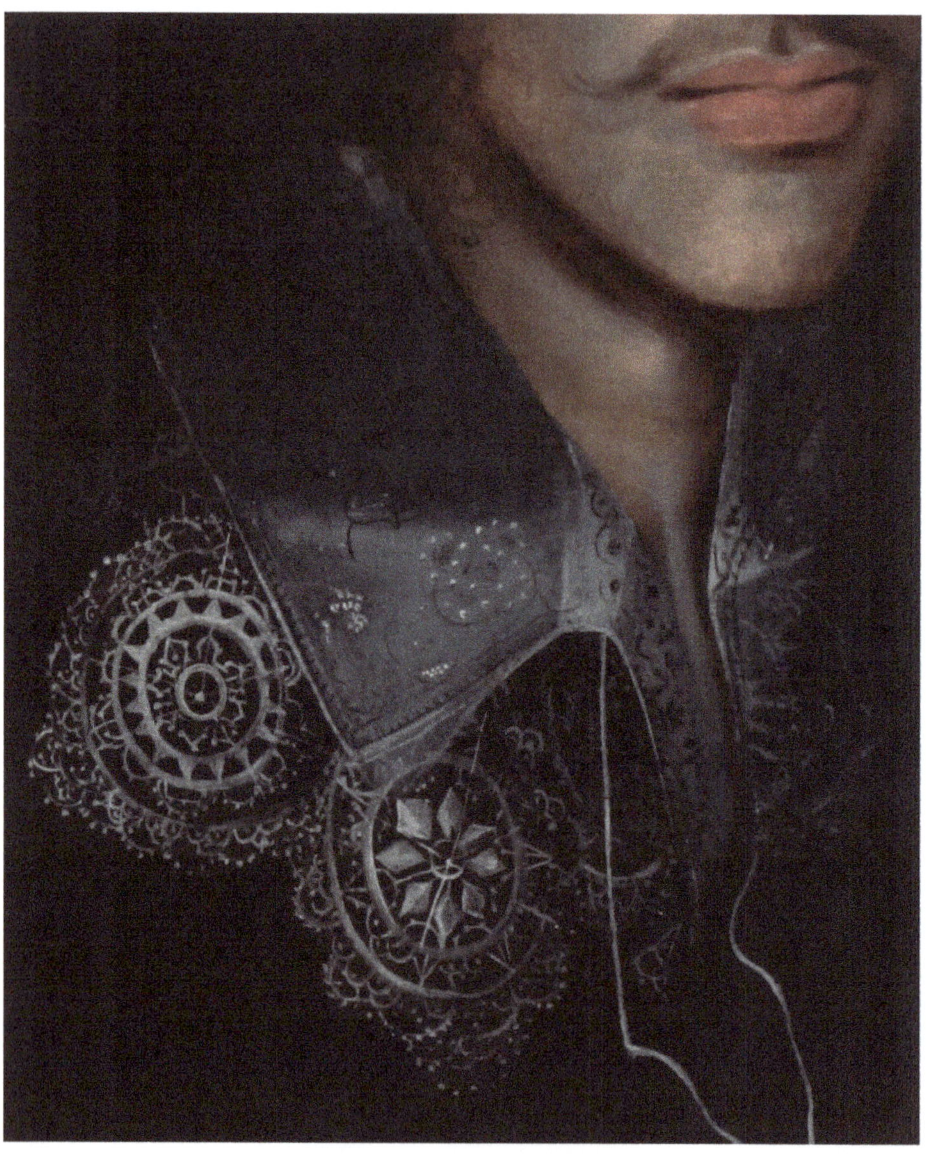

FIGURE 7.2: Detail of ... John Donne, by an unknown English artist oil on panel, circa 1595 copyright National Portrait Gallery, London (NPG 6790) Chris Titmus / Hamilton Kerr Institute, University of Cambridge.

which aptly joyned together, make one word that doth express their natures.[59]
Thus, the metal quicksilver and the planet Mercury were 'in tune', copper took 'the pattern of' Venus and lead was joined with Saturn. Since, astrologically, Mercury had connections with communication and exchange, Venus with women and love, and Saturn with age and melancholy, for example, the possibility arose for connections between metals and human states, attributes or modes of interaction.

The materials with which John Donne's second portrait were made included a pigment that was derived from lead, the Saturnine metal. In fact, because it was the product of adversity (buried in dung for a month), the pigment could be seen as displaying Saturn's 'true colours'. Lead white was used in the flesh paint (with addition of red vermilion and earths) and in the collar, the shirt-string and the lace (pure, and with varying additions of bone black) (FIGURE 7.2). In these paint passages, no inherent meaning can be ascribed to the pigment. Indeed, as a painter's notebook of the 1590s suggests, lead white was the obvious material to use, and no other material was as suitable for painting in oil.[60]

However, although not intended as a reference to an artist's pigment, the art historian Ernst Gombrich recognised that, under Neo-Platonic influence, any one thing was

> linked through the network of correspondences and sympathies with the supra-celestial essence which it embodies, it is only consistent to expect it to partake not only of the 'meaning' and 'effect' of what it represents but to become interchangeable with it.'[61]

Lead white could therefore 'embody' the 'essence' of Saturn and could even become 'interchangeable with it'. So, if Donne was to use the material nature of pigments to make puns in his portrait, it would merely be a witty extension of the everyday understanding of the world's interconnected nature that was 'the common property of every third-rate mind of the age'.[62]

Whilst the use of lead white to paint collars, shirt-strings and lace is quite unremarkable, the portrait features another use of the pigment which is specifically connected to love-sickness. Love, as we all know, is caused by Cupid's arrows. In

---

[59] Sir Thomas Browne, *Religio Medici*, (II, 2), ed. S. Greenblatt and R. Targoff (New York: New York Review of Books, 2012), p. 67.

[60] J. Murrell, 'John Guillim's book', *The Walpole Society* 57 (1993–94): p. 11.

[61] E. H. Gombrich, 'Icones Symbolicate, the visual image in Neo-Platonic thought', *Journal of the Warburg and Courtauld Institutes* 11 (1948): p. 176.

[62] E. M. W. Tillyard, *The Elizabethan World Picture* (London: Chatto and Windus, 1960), p. 101.

the circumstances surrounding this portrait, Donne's love was not reciprocated, and Donne would have known the exact nature of the arrow that had pierced his heart. Cupid's arrows were of two types. One type had a tip made of gold and the consequence of being hit by these arrows was true love, as Shakespeare implied in a play written around the time Donne's portrait was painted.[63] However, the other type of arrow had a tip made of lead and the consequence of being hit by these, according to Ovid, was unrequited love.[64]

Donne's predicament, his melancholy love-sickness, was caused by Cupid's leaden arrow. He felt a connection with a woman who was impervious to his charms and who, according to the inscription, had cast him into darkness, which is the colour of lead. (He also knew that lead's darkness was lightened by being humbled in a dung-heap. This analogy would offer the kind of double-edged interpretation in which Donne revelled – his own humbling rejection and his aloof Lady's likeness to dung.)

The consequence of Donne's unrequited love was a melancholic soul. With no established visual tradition to guide him, the unknown painter could have used any material and any form to depict Donne's ineffectively advancing soul. However, choosing – or, more likely, being told – to use lead white allowed the painting material and the painted form to enjoy a profound resonance. The melancholy state of what was depicted is implied by the melancholy nature of the material with which it was depicted. This 'material pun' was a socially and cosmologically informed, as well as a masterfully obscure, conceit typical of the young Donne's poetic output. It was reinforced by a common word that described the state of the soul. In contemporary usage, the word 'mettle' meant the 'stuff' from which a person was made, their character, disposition or temperament.[65] (Given that Donne understood his soul as a mixed and fluid entity, it is worth noting that 'temperament' shares the same etymological root as 'tempera' and 'distemper' which describe that other mixed and fluid entity, paint.) In his portrait, John Donne's melancholy mettle was made from Saturn's melancholy metal.

The portrait revisited

Later in life Donne's melancholy portrait was returned to him, presumably upon Christopher Brooke's death. By that time, Donne had recovered from the dismissal and imprisonment that resulted from his marriage to Anne. Between 1603 and 1611,

---

[63] *A Midsummer Night's Dream* (I, i, 170–72).

[64] Ovid, *Metamorphoses* (I, 470) trans. M. M. Innes (Harmondsworth: Penguin, 1975), 41.

[65] *Twelfth Night* (III, iv, 265).

John and Anne had seven children and he travelled widely in pursuit of suitable employment. In 1615 he was ordained as James I's chaplain and his future seemed assured. But his happiness was short-lived as, tragically, Anne died from complications of childbirth in 1617. John Donne died sixteen years later, spending the last ten years of his life as Dean of St Pauls, a respected member of the establishment.

Donne kept the portrait for the rest of his life and bequeathed it to a friend. He described it in his will as the portrait 'in shadows … made very many years before I was of this profession'.[66] John Donne made a sharp distinction between his early life as a theatre-going, poetry-writing privateer and his later life as a sober cleric. His friend Ben Jonson said that 'since he was made Doctor, [Donne] repenteth highly, and seeketh to destroy all his poems'.[67] Donne even suggested dual identities of the younger 'Jack Donne', the man depicted in the 'shadow' portrait, and older 'Doctor Donne', the respectable divine, Dean of St Pauls.[68] However, Donne may have been playing up the difference for rhetorical reasons when preaching at the Inns of Court where he had once been a rake.[69] Certainly, around the same time that he wrote *Love's Alchemy* he also wrote that 'in the Bible some can find out alchemy'.[70]

Donne's series of rather unconventionally-painted portraits has been described as 'an autobiographical narrative of multiple conversions; from defiant young Catholic gentleman, through radical scepticism, to professional commitment … and ultimately to an asceticism'.[71] However, there are also numerous constants running through his life, and the East-facing, death-bed shroud-painting shows, for example, a continued interest in orientation with the cosmos. He also maintained his interest in 'lower astrology', or alchemy, right until the very end.[72]

It follows that, from an astrological perspective, Donne's melancholy need not have been the passing love-sickness associated with some transient conjunction of bodies whether heavenly, carnal or both. It could also reflect his life-long

---

[66] Keynes, *Bibliography*, p. 373.

[67] *Conversations of Ben Jonson with William Drummond of Hawthornden*, as quoted in *Poems of John Donne*, ed. E. K. Chambers, vol. II, (London: Bullen, 1901), p. 240

[68] John Donne, *Letters to Several Persons of Honour* (Delmar: Scholars Facsimiles and Reprints, 1997), pp. 218, 21–22.

[69] E. Rhatigan, "The sinful history of mine own youth': John Donne preaches at Lincoln's Inn', in J. E. Archer, E. Goldring and S. Knight, eds., *The Intellectual and Cultural world of the Early Modern Inns of Court* (Manchester: Manchester University Press, 2011), pp. 90–106.

[70] *A Valediction: Of the Book*, in Donne, *Poems*, p. 18.

[71] A. Patterson, 'Donne in Shadows; pictures and politics', *John Donne Journal* 16 (1997): p. 6.

[72] J. R. Keller, 'The Science of Salvation; Spiritual Alchemy in Donne's Final Sermon', *The Sixteenth Century Journal* 23, no. 3 (1992): pp. 486–93.

relationship with Saturn, of which the youthful, passionate, love-sickness was merely one expression. (Contemporary medical theory held that there were three classes of humoural imbalance - 'mania' or violent, 'frenzy' or acute and 'melancholy' or chronic. Melancholy could be erotic, artistic or scholarly.[73]) Donne's own interpretation of the portrait 'in shadows' could have been tempered by time. It may have been a self-indulgent in-joke in his mid-twenties, but in his mid-fifties it could have become an opportunity for serious meditation. Certainly, melancholy was still a fashionable subject: Burton's 1621 *Anatomy of Melancholy* was republished in an enlarged edition in 1629, the same year that Ben Johnson's *The New Inn*, which featured love-sickness, was performed at Blackfriars.

Melancholy, being the cold and dry humour, was associated with element earth. As well as being a shadow, the background pigment 'umber' was an earth, and for Donne, earth itself was Janus-faced. In a sermon of 1624 on Paul's conversion he focused on the word 'earth' which was first associated with sin, but became transformed by its role in burial into an incubator for the soul awaiting Judgment.[74] Just as an individual pigment could be Janus-faced, so too could be the whole portrait.

The painting's inscription, 'Lighten my darkness, our Lady' is ambiguous and, true to Donne's poetic style, the Lady's identity is open to question. Donne was raised in a Catholic family and had suffered persecution, so the Lady is not likely to be the Virgin. Donne was ambitious, and aware of the dangers of inappropriate symbolism, so she is unlikely to have been Queen Elizabeth. Whilst he was widely read and knew the traditions of love poetry, the Lady is not likely to be modelled on Dante's Beatrice or Boethius' Philosophy since he chose to be depicted in the grip of sensuous passion. And Donne did not consider the soul extended between bodies as an exclusively sexual phenomenon, it was a common Elizabethan trope for friendship.[75]

Later in life, John Donne could have interpreted the Lady of the inscription as his lost wife Anne. Certainly, in an earlier poem, which bemoaned the necessary parting of lovers he described the continued connection, via souls 'gone out', between people whose bodies were at a distance.

> 'Our two souls therefore, which are one,
> Though I must go, endure not

---

[73] A. Gowland, 'The problem of early-modern melancholy', *Past and Present* 191 (2006): pp. 86–87.

[74] *Sermons*, 6:213, as quoted in D. Trevor, 'John Donne and Scholarly melancholy', *Studies in English Literature, 1500–1900* 40, no. 1 (2000): p. 94.

[75] L. J. Mills, *One soul in bodies twain; friendship in Tudor Literature and Stuart drama* (Bloomington: Principia Press, 1937).

> A breach, but an expansion,
> Like gold to aery thinness beat'[76]

Incidentally, the last line's imagery of the expanded soul's diaphanous nature draws upon the technical production of gold leaf that was used in paintings and was never seen outside the walls of a workshop. This supports the assumption that the young John Donne was aware of artists' materials and techniques.

After Anne's death, the older John doubtless mourned. But the sentiments expressed in the above poem – *A Valediction; Forbidding Mourning* – may still have rung true. The line 'Though I must go', would have to become 'Though you have gone' but he may still have felt an 'expansion' rather than a total 'breach' of their souls. Such a reading of the portrait's depicted soul would be in keeping with the Neo-Platonic image of the ascending soul as a woman which certainly influenced literature with which Donne was familiar.[77] And the soul's posthumous state was something upon which Donne speculated. In his 1601 *Metempsychosis*, he summarized the doctrine of reincarnation; 'you must not grudge to find the same soul in an Emperor, in a Post-Horse, and in a Mushroom'.[78] If a mushroom had a soul, then why not also a pigment? In which case, lead white's would be melancholy.

In the absence of textual evidence, exactly how a sitter may have perceived their portrait at any given time must, of course, remain a matter for speculation. However, what is certain is that pre-modern paintings were appreciated for their material, as well as their visible, attributes.

Amulets and talismans

Distinctions between *designo* and *colore* started to be made in the renaissance, with design (the formal aspect of art) being superior to colour (the material aspect of art).[79] However, the traditional sciences recognised that Aristotelian form and matter were mutually interdependent and both were equally necessary. Traditionally, materials had inherent powers. Just as metals' properties were derived from, or were shadows of, the planets, so stones derived their properties from the fixed stars.[80]

---

[76] *A Valediction; Forbidding Mourning*, in Donne, *Poems*, p. 33.

[77] M. H. Burcombe, 'Faire Florimell as Faire Game', *College Language Association Journal* 28, no. 2 (1984): pp. 164–75.

[78] *The Progress of the Soul, Infinitati Sacrum*, in Donne, *Poems*, p. 227.

[79] J. Gage, *Colour and Culture* (London: Thames and Hudson, 1993), pp. 171–38.

[80] Albertus Magnus, *Book of Minerals*, (II, i, 4), trans. D. Wyckoff (Oxford: Clarendon, 1967), pp. 64–67.

Appropriate stones were worn as amulets to provide protection by virtue of the material's properties and some stones had multiple properties. Lapis lazuli, for example lowered fevers, cleansed the eye and, as 'a great enemy of black choler' according to Robert Burton, 'frees the mind' to dispel melancholy.[81] Sigils were stones that had designs or images, whether natural, like fossils, or artificial, like cameos.[82] Astrological sigils were stones engraved with zodiac signs and to convey specific amuletic and talismanic benefits due to the materials of which they were composed as well as the designs they incorporated.[83]

Paintings incorporated a wide variety of materials, of which this chapter has mentioned just lead white and umber. An Italian artists' manual, translated into English in 1595, asserted that 'colours have different qualities, therefore they cause diverse effects in the beholders …(as Aristotle teacheth)'.[84] According to this treatise on art, published around the time the portrait was painted, John Donne's material pun could materially affect the viewer. The presence of a melancholy metal may have assisted him to indulge in his love-sick melancholy. And according to a treatise on melancholy published around the time the portrait was returned to him, the presence of lapis lazuli – ground up as the blue pigment, ultramarine – would have dispelled melancholy, frustrating any desire to indulge in love-sickness. Analysis of the painting shows that ultramarine was not present.

Lead white's use to depict the soul is entirely consistent with, and could contribute to, Donne's profound meditation on melancholy. In connection with another seventeenth-century metaphysical poet, it has been said that 'the affective experiences poetry offers can be exploited to gain, without readers' awareness of the goal, intuitions usually called philosophical.'[85] I would suggest that the affective experiences of 'silent poetry' can likewise bring philosophical intuitions to those who contemplate paintings since, in the pre-modern Neo-Platonic world, all material beings were infused with spirit, and astrological 'influence' was not limited to sentient beings.

---

[81] Burton, *Melancholy*, (II, iii, 4), (Amsterdam: Theatrum Orbis Terrarum, 1971), p. 442.

[82] Albertus, *Minerals* (II, iii, 1–3), p. 127–37.

[83] Albertus, *Minerals* (II, iii, 4–5), p. 138–45.

[84] Lomazzo, *Tracte* (III, 11), p. 112.

[85] R. L. Colie, *My echoing song, Andrew Marvell's poetry of criticism* (Princeton: Princeton University Press, 1970), p. 304.

PICTURING THE PRACTITIONER: TOWARDS AN ICONOGRAPHY OF ASTROLOGICAL PRACTICE

Richard Dunn

ABSTRACT: In the preface to his *Monas Hieroglyphica* of 1564, John Dee offered up an evocative image. The powerful symbolic language his text described would, he claimed, reform the observational practices of the 'astronomus' (astrologer/astronomer), who would, 'regret all the sleepless vigils and cold labors he has suffered under the open sky, when here, without any discomfort from the air, under his own roof, with windows and doors shut on all sides, at any given time, he is able to observe the movements of the heavenly bodies? And, indeed, without any mechanical instruments made from wood or brass?' Dee's textual image is the inspiration for this chapter, which will explore the possibility of using visual and literary iconography as evidence of astrological practice. Focusing on European representations of astrologers and the attributes of their work in the period 1550–1800, it will examine the extent to which such images might allow the historian to think about the practices of astrology and the relationship of astrologers to their apparatus and clients.

This chapter explores the possibility of using visual and literary iconography as evidence of astrological practice in the early modern period. It focuses on European representations of astrologers and the attributes of their work between the mid-sixteenth and early eighteenth centuries, and on surviving evidence of the practices of English consultants including Richard Napier, Simon Forman and William Lilly. Through a comparison and discussion of these and other sources, it will examine the extent to which visual and literary images might allow the historian to think about the practices of astrology and the relationship of astrologers with their clients as mediated by the spaces of practice and the tools (broadly conceived) that they used.

In exploring these questions, it is important to recognise some of the issues arising. One concerns terminology and identification. In a period when terms such as astrologer, astronomer, magus or philosopher could be used in a range of sometimes ambiguous, sometimes interchangeable, ways, one could easily ask, when is an astrologer an astrologer? For the sake of simplicity, the images discussed in this chapter focus on those specifically identifying their subject as an astrological practitioner, or where the image is part of an astrological text (e.g., its frontispiece). There are clearly striking similarities with contemporary images of astronomers, such as Johannes Vermeer's 'The Astronomer' and Gerrit Dou's 'Astronomer by Candlelight', but a discussion of these would require a longer exposition than is possible in this chapter.[1]

---

[1] Johannes Vermeer, 'The Astronomer', oil on canvas, 1668 (Musée du Louvre, 1983–28); Gerrit

A second issue concerns the different forms of astrological practice one might identify. Loosely following Patrick Curry's division of astrology into categories such as natural or judicial (itself comprising several modes) and popular or high/elite, the practice of astrology might range from tasks such as the composition of almanacs and manuals, through receiving or visiting clients and giving them advice, to philosophical speculation and writing about the nature and operation of astrological influences.[2] What follows will predominantly think about the client-based practices of judicial astrology, using English consultants of the seventeenth century as the main examples.

Setting the tone: Iconographic considerations

A useful image with which to open the discussion is a depiction entitled 'Der Astronomus' (FIGURE 8.1). This appeared in the German poet Hans Sachs's *Eygentliche Beschreibung Aller Stände auff Erden* (Exact Description of All Ranks on Earth) (Frankfurt, 1568) and in Hartmann Schopper's Latin edition of the work, *Panoplia omnium illiberalium mechanicarum* (Book of Trades) (Frankfurt, 1568), both of which were illustrated with woodcuts by a Swiss printmaker, Jost Amman. The two works feature a series of illustrations of different occupations, each accompanied by a short description in verse, with the whole arranged according to a 'natural' social hierarchy. In each case, the importance of hard work and modest living are emphasised as routes to a pious existence and a harmonious society: these are essentially moral texts.[3]

For the 'Astronomus', the inscription that accompanies the image makes it clear that this is a man whose trade is astrological forecasting. He can 'predict eclipses / By reading the sun, moon and stars', enabling him to forecast whether 'the coming year will be fruitful, / Or whether increased prices and the dangers of war / Or other misfortunes are to be expected'. The illustration itself shows a lone figure in a study, using a pair of dividers to measure points on a globe, surrounded by other globes and time-telling instruments, including two sundials on the windowsill. In its depiction of solitary, contemplative activity, Amman's image seems to draw most explicitly on representations of St Jerome, the iconography of which was well established

---

Dou, 'Astronomer by Candlelight', oil on panel, late 1650s (The J. Paul Getty Museum, 86.PB.732).

[2] Patrick Curry, *Prophecy and Power: Astrology in Early Modern England* (Cambridge: Polity Press, 1989), pp. 7–15.

[3] Alison M. Kettering, 'Men at Work in Dutch Art, or Keeping One's Nose to the Grindstone', *The Art Bulletin* 89, no. 4 (2007): pp. 694–714.

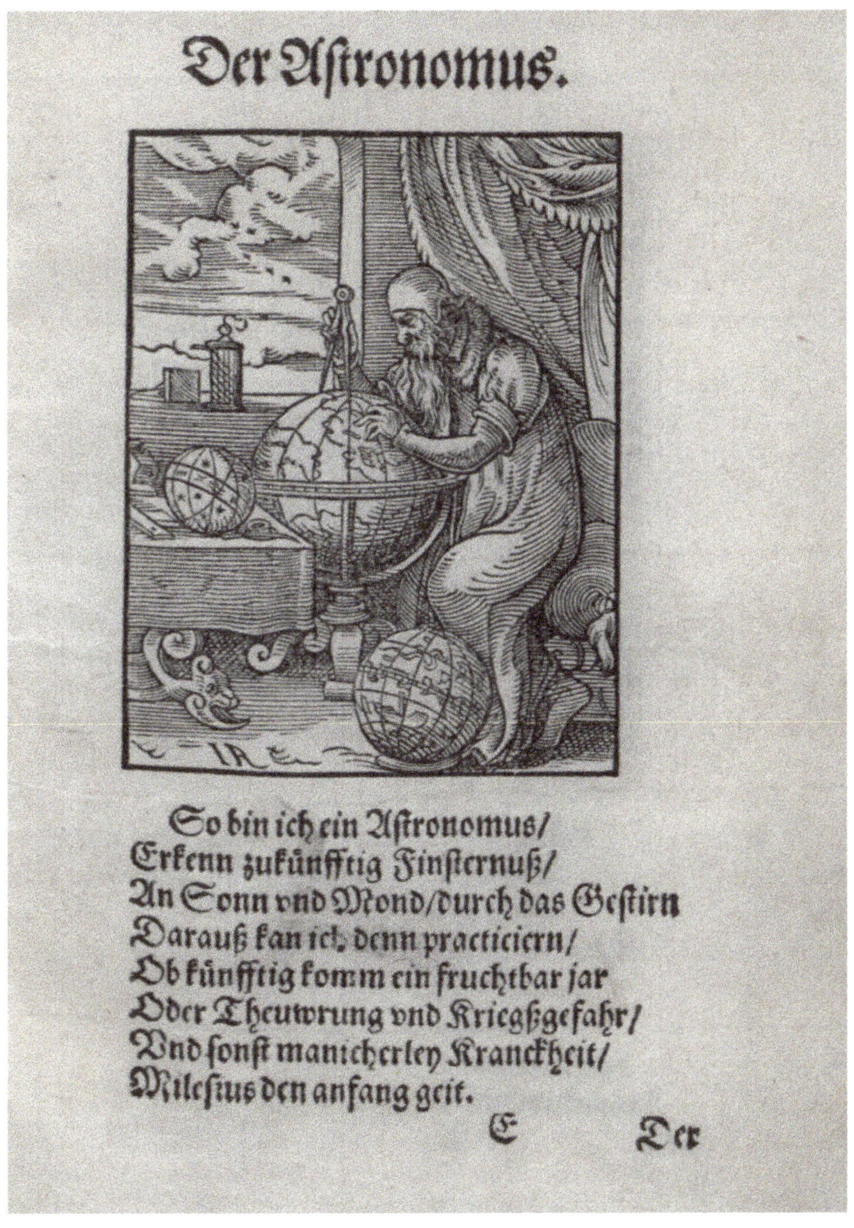

FIGURE 8.1: Jost Amman, 'Astronomus', from Hans Sachs, *Eygentliche Beschreibung Aller Stände auff Erden* (Frankfurt, 1568) (Saxon State and University Library Dresden (SLUB)). Photo: SLUB/Deutsche Fotothek.

by the sixteenth century and became the model for later illustrations of scholars of all sorts in their studies, either locked away from sight or with a window onto the outside world.[4]

The instruments Amman depicts draw on additional iconographies and are significant in symbolic terms. The dividers (which are sometimes identified as compasses) can be linked, among other things, to the disciplines of astronomy and geometry, to justice and prudence, and to acts of creation.[5] Globes can simply be a symbol of the heavens, but can also have richer and more ambiguous meanings, as Catherine Hofmann has explained:

> On the one hand, the rotating sphere evokes the fickleness of love and the reverses of Fortune, but also the world doomed to impermanence and death, along with the vanity of all earthly things. On the other, as the perfect geometric figure, the globe evokes the power of the mighty or the knowledge of the learned, but also the ultimate heavenly refuge, the firmament that encompasses everything and is supposedly perfect and immutable.[6]

In addition, a point that will be relevant later, a sphere held in the hand may offer the reading that free will and wisdom can counteract or overcome the influence of the stars or fate. In the same manner, 'fato prudentia maior' (prudence is greater than fate) was a common motto in books of emblems and other works.[7]

The instruments depicted by Amman for his astrologer are, unsurprisingly, those that also typically appear in images of Astrologia or Astronomia as muses or personifications of one of the seven liberal arts. An image from such a series by Georg Pencz in 1541, for example, shows Astrologia and a putto with a globe, dividers and a quadrant (an astronomical observing instrument); a similar image published by Hieronymous Cock in 1565 has a globe, dividers and sundials.[8] What Amman's 'Astronomus' seems to draw on, then, is an established set of attributes

---

[4] A. Jasińska, 'The Portrait of a Scholar in Painting. An Outline of the History and Development of Iconographic Types', in A. Jasińska and E. Wyka, eds, *Uczony I Jego Pracownia. The Scholar and his Study* (Kraków: Jagiellonian University Museum, 2005), pp. 9–23.

[5] James Hall, *Dictionary of Subjects and Symbols in Art* (Revised edition, London: John Murray, 1979), p. 73; Juan Eduardo Clirlot, *A Dictionary of Symbols* (Mineola, NY: Dover Pub., 2002), p. 61.

[6] Catherine Hofmann, 'The Globe as Symbol in Emblem Books in the West, Sixteenth and Seventeenth Centuries', *Globe Studies* 49/50 (2002): pp. 81–120 (quote pp. 119–20).

[7] Hofmann, 'The Globe as Symbol', p. 105.

[8] Georg Pencz, 'Astrologia', from 'The liberal arts', engraving, about 1541 (British Museum, E,4.280); Cornelis Cort, after Frans Floris, 'Astrologia', from 'Seven Liberal Arts', engraving, published by Hieronymous Cock, 1565 (British Museum, F,1.292).

for the representation of astrology or astronomy and for the scholar as a solitary and contemplative figure.

The main question for this chapter, however, is whether images such as this can tell us something meaningful about routine astrological practice in the early modern period. In exploring this, I am particularly interested in thinking about the practice of consultant astrologers and their relationships and interactions with their clients. Amman's astrologer is not a consultant. Rather, he is identified in the accompanying verse as the author of predictions found in printed almanacs. Images of astrological practitioners one might identify more specifically as consultants do appear, however, in other books of trades in the same vein as Sachs's. Christoph Weigel's 1709 work, *Centi-folium stultorum in quarto, oder, Hundert ausbündige Narren in folio*, for instance, shows an astrologer and client in discussion (FIGURE 8.2). A woman, accompanied by two children, is shown paying the astrological consultant, who points to the sky, where zodiacal and planetary symbols appear. Behind the astrologer are the tools of the trade: books (open on the table as well as in the bookcase); a celestial globe; dividers; a telescope; and what appears to be a lens. These items hint at the different activities on which the astrologer's practice draws: observation (telescope); representation and measurement (globe and dividers); codification and calculation (books). Strikingly similar is another image of an astrologer with a client, from Robert Fludd's *Utriusque cosmi maioris scilicet et minoris metaphysica, physica atque technica historia*, a long, dense and copiously illustrated philosophical discussion of the macroscosm and microcosm (FIGURE 8.3). Again, the celestial globe and dividers are placed prominently, this time between the client and the astrologer, who is drawing up an astrological figure, while what appears to be an astronomical or astrological text is open on a table behind the client, with a pair of spectacles next to it. Further books can be seen on the shelf high up on the wall above the astrologer. The scene is imagined, but again clearly draws on recgonisable iconographies.

Identity and image: Seventeenth-century frontispieces

Switching to images of real astrologers, the most obvious source is portraits produced as frontispieces. As Inga Elmqvist Söderlund notes, frontispieces and title page illustrations, which might be a book's only printed image in this period, act as advertisements, signalling to the potential buyer or reader what kind of book it is.[9] According to Bernard Bauhuis, writing in 1617, for instance, 'At the beginning of the book ... many people would like to see some engraving ... It amuses the reader

---

[9] Inga Elmqvist Söderlund, *Taking Possession of Astronomy: Frontispieces and Illustrated Title Pag-*

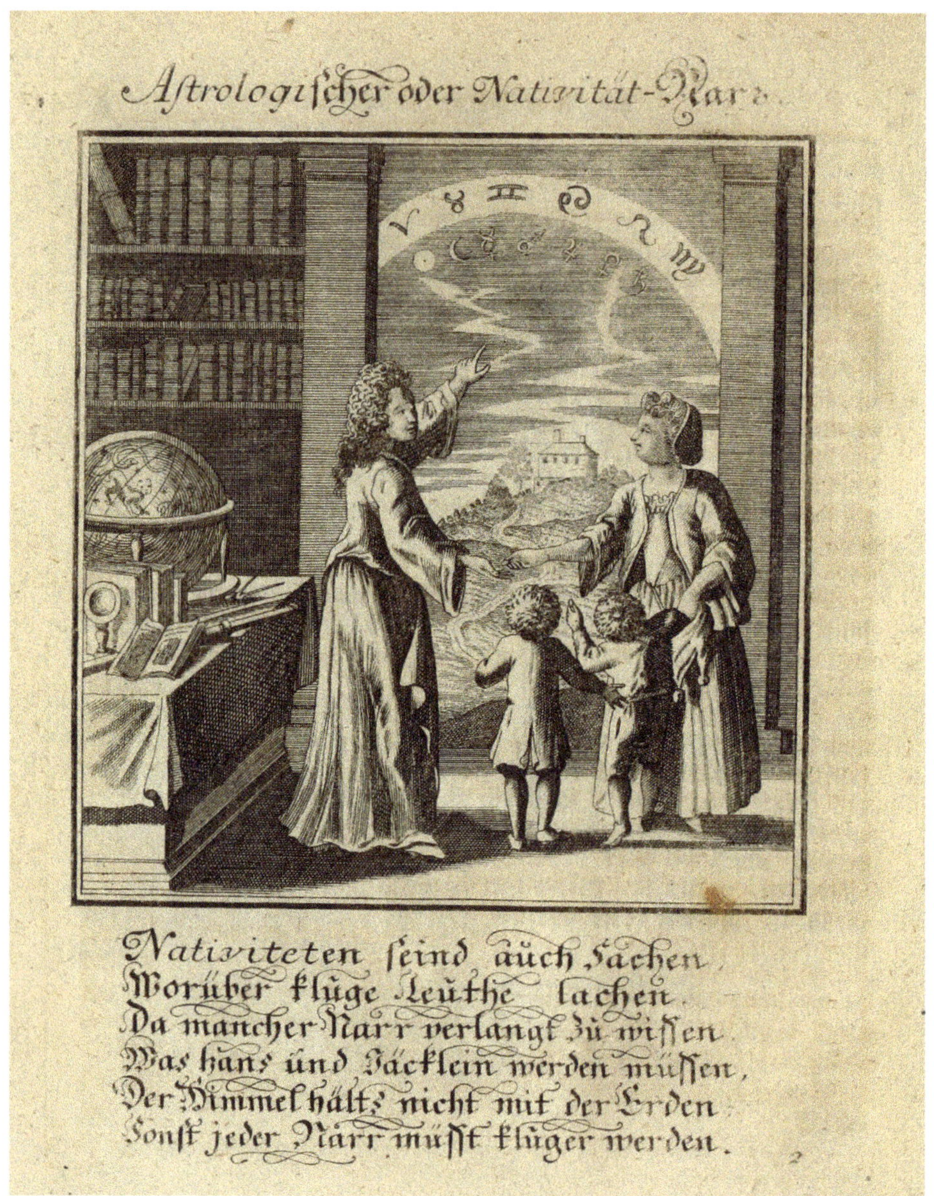

FIGURE 8.2: 'Astrologischer oder Nativität-Narr', from Christoph Weigel, *Centi-folium stultorum in quarto, oder, Hundert ausbündige Narren in Folio* (Vienna: Johann Carl Megerle; Nuremberg: Johann Christoph Weigel, 1709) (Getty Research Institute).

146

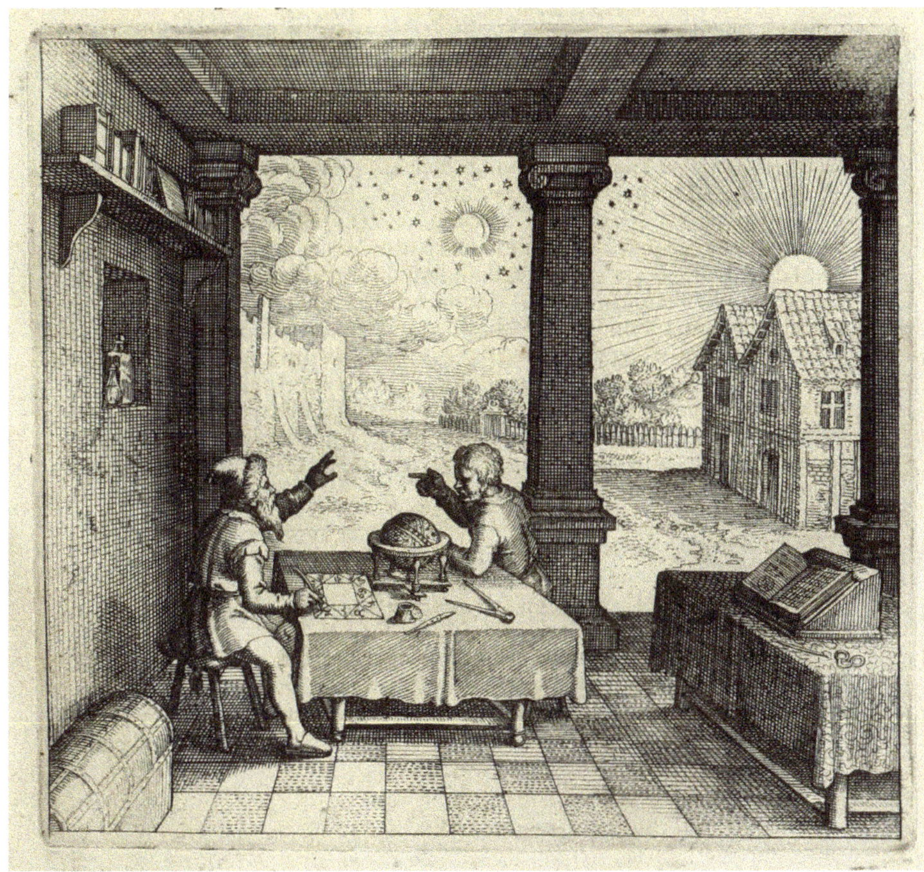

FIGURE 8.3: Robert Fludd, *Utriusque cosmi maioris scilicet et minoris metaphysica, physica atque technica historia*, 2 vols (Oppenheim: Johann Theodor de Bry, 1617–21), 2: Book I, Section II, Part IV, p. 71 (Getty Research Institute).

wonderfully, it attracts the buyer, it decorates the book and it does not add much to the price'.[10] While the frontispiece might be able to do no more than hint at the book's detailed content, it is the place to establish the author's credentials, assuring the potential reader of the writer's legitimacy as an authority on the subject at hand. This certainly appears to be the case with frontispiece portraits of English astrologers.

---

es in 17th-Century Books on Astronomy (Stockholm: Center for History of Science, Royal Swedish Academy of Sciences, 2010), pp. 88, 313.

[10] Bernard Bauhuis to B. Moretus, 1 August 1617, quoted in Söderlund, *Taking Possession of Astronomy*, p. 89.

A good example of this genre is the portrait of William Lilly (1602–1681) (FIGURE 8.4), which appeared as the frontispiece to his seminal *Christian Astrology* (1647), the first major astrological textbook in the English language.[11] Lilly appears in the clothes of a gentleman and is accompanied by familiar attributes: the celestial globe on which his right hand rests; a printed text showing the symbols for the signs, planets and aspects; writing implements (a pen and inkstand); and an astrological figure in his left hand. Through the window is a more obviously symbolic scene: Saturn and Jupiter shown in conjunction near the cusp of Pisces and Aries, above a pastoral scene that includes a shepherd and a fisherman. As Catherine Blackledge explains in her biography of Lilly, this background image draws on known aspects of his life. The zodiacal positions of Jupiter and Saturn are those they occupied at their most recent conjunction in 1643, from which Lilly had predicted the downfall of King Charles I. Their positions also reflect the then current suggestion that the birth of Christ had been heralded by a conjunction of Saturn and Jupiter on the cusp of the same two signs. The pastoral scene below alludes to Lilly's childhood as a farmer's son in Lincolnshire (although he lacked any agricultural bent) and to a straightforward Christian reading of the two figures as 'Fisher of Souls' and 'Shepherd of Men'.[12]

As Blackledge notes, the astrological figure in Lilly's left hand is also significant. It has the representation of the twelve houses as triangles within a square, the conventional form for an astrological chart or figure in this period, although it lacks the numerical indications and symbols for the house divisions and planetary positions. Lilly's age ('Aetatis 45') is written at the top, while the centre of the figure has the words 'Non cogunt' (they do not compel) written across it. According to Curry, these words appear in every extant portrait of Lilly and were hugely significant to his position on astrological determinism. The idea that the stars incline but do not compel was a crucial maxim throughout Lilly's astrological writings. This was in contrast to the hard-line beliefs of a few of his contemporaries: the frontispiece to Elias Ashmole's *Fasiculus Chemicus* (1650), for example, includes a figure for Ashmole's nativity with the inscription 'Astra regunt homines' – the stars rule men – at its centre.[13] Lilly's less emphatic assertion, with which the majority of astrologers at this time

---

[11] Patrick Curry, 'Lilly, William (1602–1681)', *Oxford Dictionary of National Biography*, Oxford University Press, 2004 (hereafter ODNB), at http://www.oxforddnb.com/view/article/16661 [accessed 28 September 2018].

[12] Catherine Blackledge, *The Man Who Saw the Future: A Biography of William Lilly* (London: Watkins, 2015), pp. 47–50.

[13] Curry, *Prophecy and Power*, p. 30.

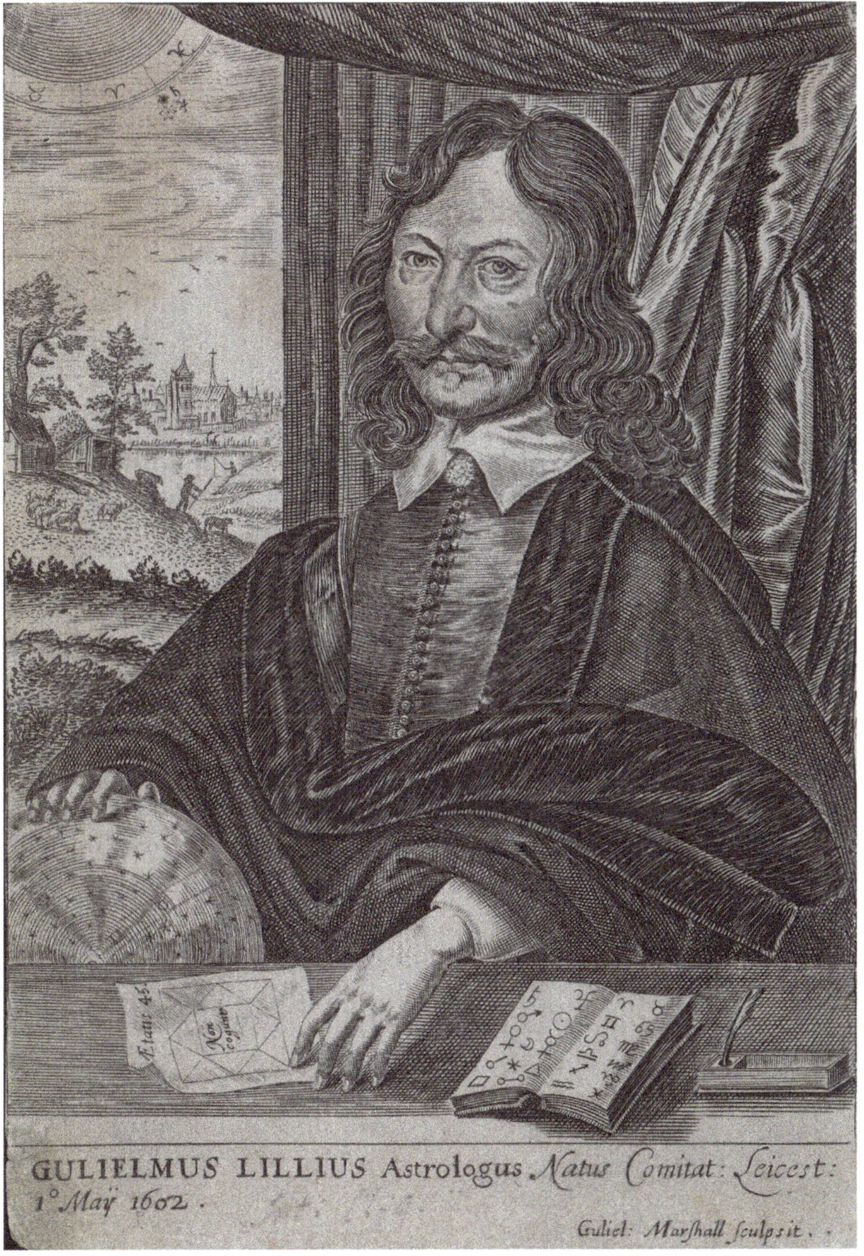

FIGURE 8.4: William Lilly, engraved by William Marshall, from *Christian Astrology* (London: Thomas Budenell, 1647), frontispiece (Private collection). Photo: National Maritime Museum, Greenwich

would have agreed, was an important point in the ongoing battle to demonstrate that the celestial art was compatible with religion and, above all, the principle of human free will on which a Christian notion of salvation depended.[14] Intriguingly, the iconography of the published portrait seems to support this reading in other ways, in particular in the placing of Lilly's right hand on the celestial globe. This seems to draw on representations of a sphere held in the hand, which, as discussed above, can convey the idea that wisdom or the wise man can overrule the stars.

A number of strikingly similar portraits appeared in other English astrological works in this period, including that of Lilly's amanuensis, Henry Coley (1633–1707), who had a flourishing practice as a consultant astrologer and teacher from his home in Baldwin's Gardens, off Gray's Inn Road in London.[15] In the image of a 35-year-old Coley from his astrological manual, *Clavis Astrologiæ; or, a Key to the whole Art of Astrologie* (1669) (FIGURE 8.5), a celestial globe and dividers are once again prominent. In this instance, however, it is the dividers rather than the globe that are in Coley's hand, allowing him to draw or measure the astrological figure on the table.

The apparatus of astrological consultations

Thinking about these images of known consultants – in the examples illustrated of Lilly and Coley, both are from manuals of the art of judicial astrology – it is interesting to speculate on what they can tell us, if anything, about their routine practices with clients. To do so, however, one must draw on other sources to think about the details of the consultation. Frontispiece portraits were not, after all, presented as realistic representations of the astrologer in his working environment.

Surviving casebooks for English astrological consultants including Lilly, Simon Forman (1552–1611), Richard Napier (1559–1634) and John Booker (1602–1667) give a few tantalising details.[16] These indicate that clients normally consulted the astrologer in person, although they could also pose their questions by letter or through a

---

[14] Ann Geneva, *Astrology and the Seventeenth Century Mind: William Lilly and the Language of the Stars* (Manchester: Manchester University Press, 1995), pp. 9–10; Keith Thomas, *Religion and the Decline of Magic* (London: Weidenfeld and Nicolson, 1971), pp. 361–64.

[15] Bernard Capp, 'Coley, Henry (1633–1704)', *ODNB*, at http://www.oxforddnb.com/view/article/5899 [accessed 28 September 2018]; Curry, *Prophecy and Power*, pp. 88–89.

[16] Jonathan Andrews, 'Napier, Richard (1559–1634)', *ODNB*, at http://www.oxforddnb.com/view/article/19763; Lauren Kassell, 'Forman, Simon (1552–1611)', *ODNB*, at http://www.oxforddnb.com/view/article/9884; Jonathan Andrews, 'Napier, Richard (1559–1634)', *ODNB*, at http://www.oxforddnb.com/view/article/19763; Bernard Capp, 'Booker, John (1602–1667)', *ODNB*, at http://www.oxforddnb.

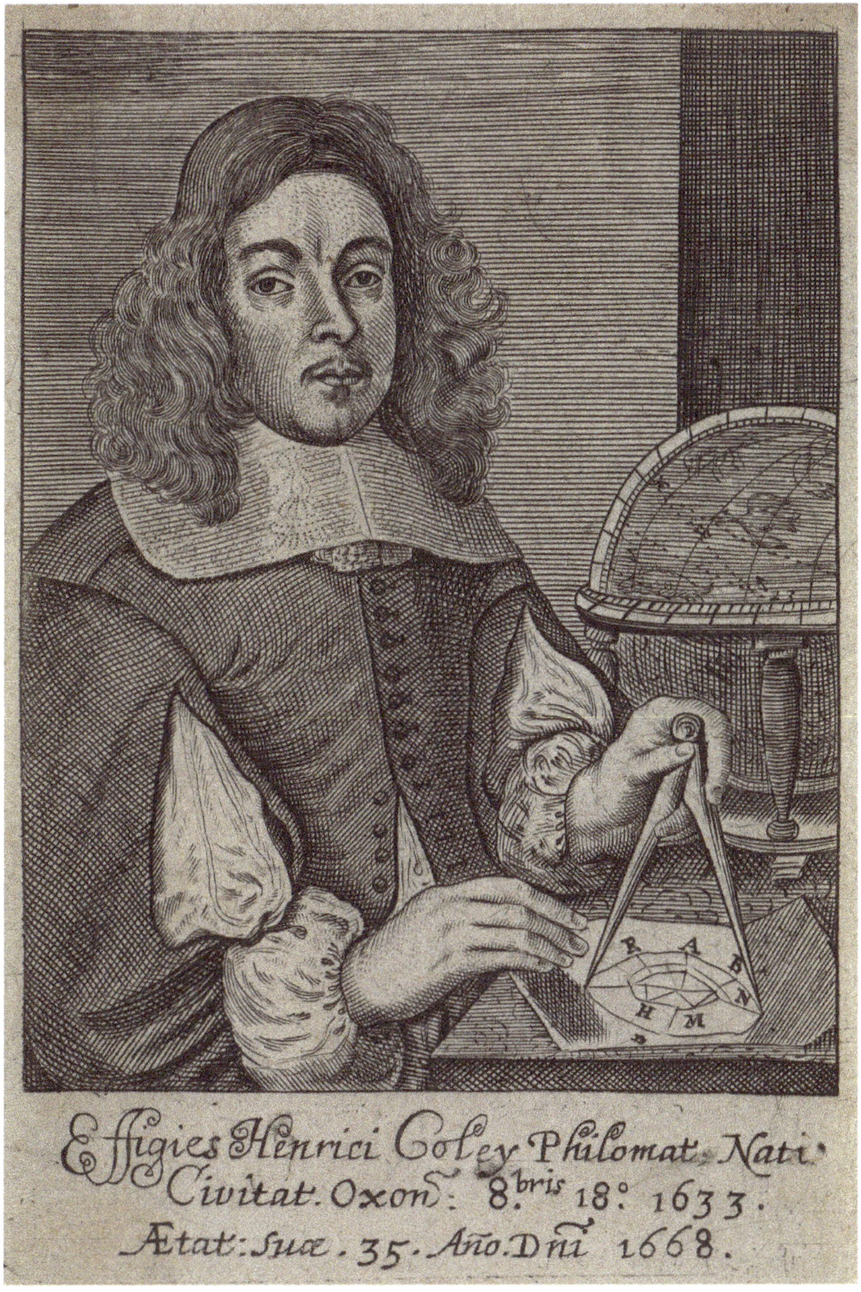

FIGURE 8.5: Henry Coley, from *Clavis Astrologiæ; or, a Key to the whole Art of Astrologie* (London: Printed for Joshua Coniers, 1669), frontispiece (Private collection). Photo: National Maritime Museum, Greenwich

messenger or other intermediary. On receipt of the question, the astrologer recorded the precise time at which it was asked, then drew up a figure for that time and gave a judgement. For an experienced consultant, the whole operation from receiving the client to offering advice or a judgment might take less than a quarter of an hour.[17] The practice was based on horary astrology, the most deterministic form of the art, but astrologers such as Lilly would carefully reiterate that their predictions were not absolutely binding – 'non cogunt', he would insist.

The consultation generally took place in a study or similar room at the astrologer's house. William Lilly had a 'Corner House on the Strand' in London, where he administered to nearly 2000 clients a year.[18] Forman occupied a number of residences in or near London during his working life, and certainly had a study for his consultations and other work. A casebook entry for 15 June 1596 records, for example, 'Halek [had sex] prius Elizabeth Hipwell ... studio', i.e. in the study.[19] Twenty years later, Forman's wife stated that he often remained 'alone with Mistress [Anne] Turner in his study for hours at a time'.[20] Lauren Kassell notes that some of Forman's consultations also took place at his clients' homes.[21] Like Forman, John Booker operated from addresses in London, while Richard Napier worked from his rectory in Great Linford in Buckinghamshire. Intriguingly, John Aubrey later wrote of Napier that, 'When a Patient, or Querent came to him, he presently went to his Closet to Pray: and told to admiration the Recovery, or Death of the Patient', suggesting, at least in Napier's case, the use of more than one space for the operation.[22]

The number of clients seen by these consultants was impressive. Forman recorded over 10,000 consultations between 1596 and 1603, and his practice had begun at least a decade earlier and continued until he died in 1611. Napier's records, which survive from the start of his astrological work in 1597 to his death in 1634, give

---

com/view/article/2865; all URLs here accessed 28 September 2018.

[17] Curry, *Prophecy and Power*, pp. 8–9; Thomas, *Religion and the Decline*, pp. 286–87, 305–7.

[18] Curry, *Prophecy and Power*, p. 29; Curry, 'Lilly, William'.

[19] Simon Forman, figure for Elizabeth Hipwell, 15 June 1596, Bodleian Library, Oxford, MS Ashmole 234, fol. 59v, at http://www.magicandmedicine.hps.cam.ac.uk/view/case/normalised/CASE443 [accessed 28 September 2018].

[20] Quoted in Judith Cook, *Dr. Simon Forman: A Most Notorious Physician* (London: Chatto & Windus, 2001), p. 194.

[21] Lauren Kassell, *Medicine and Magic in Elizabethan London: Simon Forman: Astrologer, Alchemist and Physician* (Oxford: Clarendon Press, 2005), p. 137.

[22] John Aubrey, *Miscellanies* (London: Printed for Edward Castle, 1696), p. 133.

details of around 70,000 consultations.[23] Booker dealt with over 16,500 enquiries in the period from 1648 to 1665.[24]

Garnering more specific details of the environment, practice and tools of the consultation is more challenging, however, since most casebook entries simply show the question asked, the figure cast and the judgement in brief; some do not even give as much detail as that.[25] In a few places, however, there are tantalising scraps of evidence. Based on his reading of the casebooks, Michael MacDonald eloquently reconstructs Napier's routine:

> Napier treated between five and fifteen patients a day, repeating with each of them a ritual of interrogation and annotation that invoked the axioms of astrological medicine. ... When the precise coordinates of the patient's identity had been recorded, the astrologer noted the time and date and drew a cross-hatched box [the figure] on the page. Deciphering the ephemeris at his elbow, he mapped the heavens at that moment, positioning the symbols of the planets and the signs of the zodiac along the celestial frame. The purpose of this astral cartography was to situate the patient in the cosmos, placing him at the vortex of the natural forces that impelled the universe, discovering the correspondences that linked microcosm and macrocosm.[26]

The main material element MacDonald identifies, therefore, is the horary figure, which was drawn up and noted in the casebook as part of the consultation (so pen, ink and paper were needed). Kassell concurs in her discussion of Simon Forman's London practice, noting that the casebook, containing the horary figures and notes drawn up each day, was the main item present at consultations. 'Astrology', she emphasises, 'was by definition a written art'.[27] MacDonald goes even further, suggesting that the key difference between the astrologer and other medical practitioners was the astrological figure itself, which not only acted as an aid to the astrologer but also had some palpable effect for the client.[28]

The other supplementary aid identified in MacDonald's reconstruction is the

---

[23] Casebooks Project, 'Introduction to the casebooks, at http://www.magicandmedicine.hps.cam. ac.uk/the-manuscripts/introduction-to-the-casebooks [accessed 28 September 2018]; Kassell, 'Forman, Simon'.

[24] Thomas, *Religion and the Decline*, p. 307.

[25] Casebooks Project, 'How to read the casebooks', at http://www.magicandmedicine.hps.cam.ac.uk/ the-manuscripts/how-to-read-the-casebooks [accessed 28 September 2018].

[26] Michael MacDonald, *Mystical Bedlam: Madness, Anxiety, and Healing in Seventeenth-Century England* (Cambridge: Cambridge University Press, 1981), p. 26.

[27] Kassell, *Medicine and Magic*, pp. 129, 132.

[28] MacDonald, *Mystical Bedlam*, p. 30.

ephemeris, the set of tables of astronomical data needed to construct the horary figure. As Lilly noted in *Christian Astrology*, 'without a present *Ephemeris* and *Table of Houses* it's impossible to instruct you to set a figure, without which we can give no judgement, or perform anything in this art'.[29] Further evidence of the use of named ephemerides can be found in the casebooks of Richard Napier and Simon Forman (who had instructed Napier in the astrological art). Napier, it seems, kept more than one ephemeris on hand and occasionally recorded the data from different authors, suggesting that there was some art in balancing or choosing between the alternative interpretations that might result from their differing values. Casting a figure for William Colles on 15 May 1599, for example, he records data for the lunar aspects from 'Everart[us]' and 'Stadius'. Likewise, 'Orig:' and 'Everart[us]' are identified as sources in January 1602.[30] These entries presumably refer to Martin Everart's *Ephemerides novae et exactae* (1597 and later editions), Johannes Stadius's, *Ephemerides novae et exactae* (1556), or perhaps his *Tabulae Bergenses* (1560), and the *Ephemerides novae* (1599) of David Origanus. In another instance, Napier's judgment seems to hinge on which ephemeris, one of which is identified as Everart, he chooses to trust in the matter of a possible pregnancy, although the alternative set of tables is not named.[31] Overall, references to Everart's ephemerides seem to be slightly more common, indicating some preference on Napier's part. Forman refers to Everart's ephemerides in several casebook entries as well. He also claimed during his examination by the College of Physicians in 1593 that he practised his art, 'using no help but the Ephemerides', perhaps supporting MacDonald's interpretation.[32]

References to instruments and their use are, however, frustratingly absent from the surviving casebooks, with Forman's evidence to the College of Physicians perhaps suggesting that instruments were of little use in his work. One notable excep-

---

[29] William Lilly, *Christian Astrology* (London: Thomas Budenell, 1647), p. 32.

[30] Richard Napier, figure for William Colles, 15 May 1599, Bodleian Library, Oxford, MS Ashmole 228, fol. 164r, at http://www.magicandmedicine.hps.cam.ac.uk/view/case/normalised/CASE12464 [accessed 28 September 2018]; Napier, figure for Mrs Garnar, 2 January 1602, Bodleian Library, Oxford, MS Ashmole 404, fol. 256v, at http://www.magicandmedicine.hps.cam.ac.uk/view/case/normalised/CASE20187 [accessed 28 October 2015].

[31] Napier, figure for Mrs Troughton's servant, 2 January 1602, Bodleian Library, Oxford, MS Ashmole 404, ff. 256v–257r, at http://www.magicandmedicine.hps.cam.ac.uk/view/case/normalised/CASE20189 [accessed 28 September 2018].

[32] For example, Simon Forman, figure for himself, 1 January 1599, Bodleian Library, Oxford, MS Ashmole 219, fol. 135r–v, at http://www.magicandmedicine.hps.cam.ac.uk/view/case/normalised/CASE5888 [accessed 28 September 2018]; Forman's evidence to the College of Physicians quoted in S. P. Cerasano, 'Philip Henslowe, Simon Forman, and the Theatrical Community of the 1590s', *Shakespeare Quarterly* 44, no. 2 (1993): pp. 145–58 (p. 146).

tion occurs in Richard Napier's casebooks. In July 1598, he cast a figure on behalf of Harry Mun, who had sent him a urine sample. Napier noted that, 'Mun sent his water to me about … 6 … but we aymed the houre because it was rayny'.[33] This suggests that Napier normally used a sundial or similar instrument for determining the time during the day (as Amman's 'Astronomus' does), but was unable to do so on this occasion and had to estimate (aim) the time instead – a victim of inclement English weather. It seems likely that most consultants would have used sundials to find the time, since they were cheaper and more reliable (weather aside) than clocks and watches until at least the end of the seventeenth century. Further research may, however, shed more light on this question.

Other modes of representation: Polemic and satire

To find additional evidence of the other aids that astrological consultants might have used or at least had in the room, one has to look further afield, and with some caution, to other types of text and image. The most striking description of an astrologer's consulting room from this period can be found in John Melton's *Astrologaster, Or, The Figure-Caster* (1620). This source must be treated carefully, however, since it is an attack on astrology and its false practitioners, as the book's subtitle makes clear:

> the arragnment of Artlesse Astrologers, and Fortune-tellers, that cheat many ignorant people under the pretence of foretelling things to come, of telling things that are past, finding out things that are lost, expounding Dreames, calculating Deaths and Nativities …

Nonetheless, the description of the consulting room is highly evocative. Having come to the house of 'Doctor P. C.' in Moorfields, whom he already believes to be a charlatan, Melton describes the study of the 'Wise-man':

> Before a Square Table, covered with a greene Carpet, on which lay a huge Booke in *Folio*, wide open, full of strange Characters, such as the Ægyptians and *Chaldeans* were never guiltie of; not farre from that, a silver Wand, A Surplus, a Watering Pot, with all the superstitious or rather fayned Instruments of his cousening Art. And to put a fairer colour on his black and foule Science, on his head hee had a foure-cornered Cap, on his back a faire Gowne (but made of a strange fashion) in his right hand he held an Astrolabe, in his left a Mathematicall Glasse.[34]

---

[33] Napier, figure for Harry (Henry) Mun, 24 July 1598, Bodleian Library, Oxford, MS Ashmole 228, fol. 4v, at http://www.magicandmedicine.hps.cam.ac.uk/view/case/normalised/CASE11572 [accessed 28 October 2015].

[34] John Melton, *Astrologaster, Or, The Figure-Caster* (London: Barnard Alsop, 1620), p. 8.

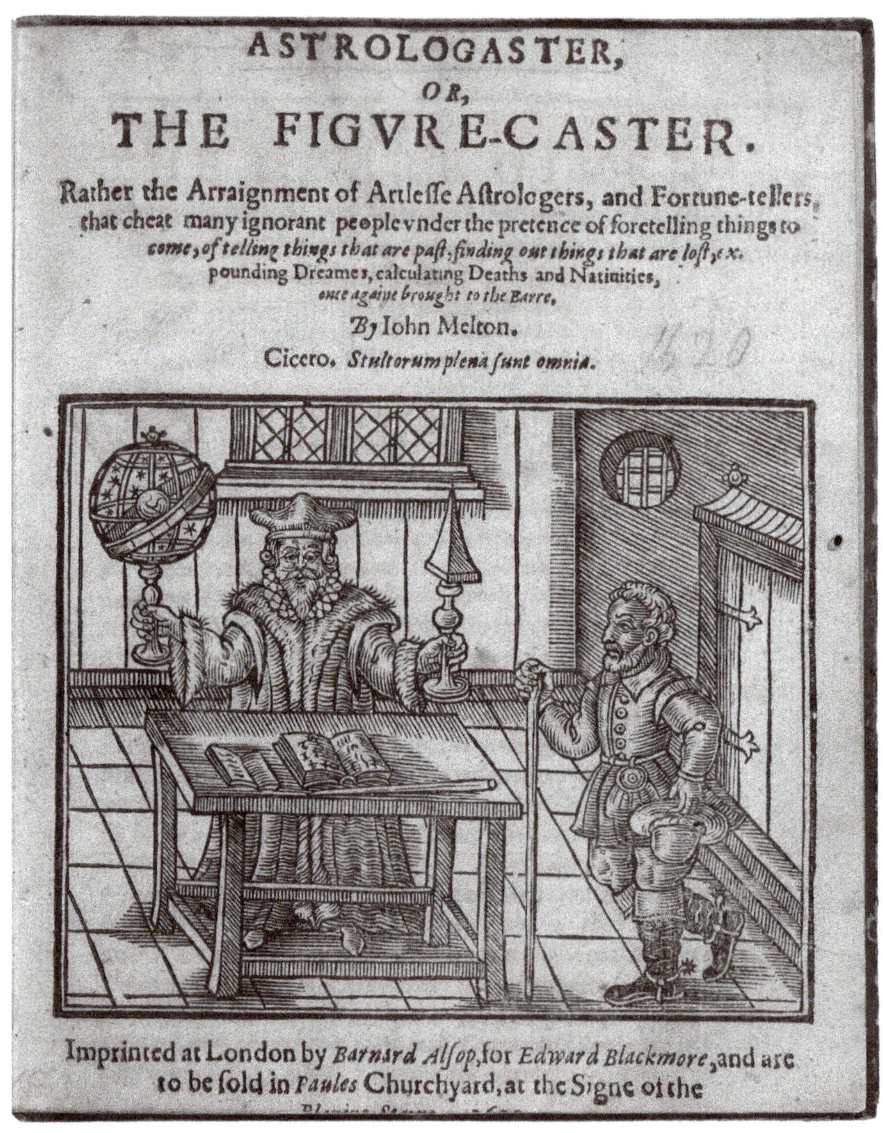

FIGURE 8.6: John Melton, *Astrologaster, Or, The Figure-Caster* (London, 1620), title page. By permission of the Folger Shakespeare Library.

Melton describes a range of aids therefore, some of them seemingly magical, such as the silver wand and watering pot, others more clearly for celestial observation or calculation, notably the astrolabe, although the author later derides all the astrologer's apparatus as 'meere superstitious Ornaments' akin to the trappings of papists (the book is vehemently anti-Catholic). The title page also illustrates the scene (FIGURE 8.6), with the astrologer shown holding the astrolabe (in fact rendered as an armillary sphere) and the 'Mathematical Glasse', an object with which he claims he can see 'all things done in Christendome'.[35] The wand sits with his books on the table. Books such as the ephemeris have their place in the text, the astrologer declaring that, 'this Booke (meaning his *Ephemerides*) with my Art and Industrie, shall be the Instrumentall Causes to make you happie in the recoverie of that which is worthie both of my Care and your Cost'.[36]

It is worth noting the resemblance of *Astrologaster*'s title page to that of Christopher Marlowe's *The Tragicall History of the Life and Death of Doctor Faustus*, which was published the same year and most likely used Forman, who was gaining some notoriety by then, as a model for its main protagonist.[37] This image shows Faustus standing inside a magical circle marked with astrological and other glyphs, with a book in his hand, while a demon stands to the right in the foreground. The image used for *Astrologaster* also subsequently appeared on the title page of a very different work, *The Dutch Fortune-Teller ... Brought into England by John Booker* (1650 and later editions). This sets out the principles of a divinatory system combining letters, words and numbers, and making use of dice, 'wheels' and 'globes' to make predictions. Booker thought of the system as a form of natural magic to be taken seriously, but complained that others seemed to see it as a jest.[38] The title-page image for *Astrologaster* seems, therefore, to have aligned easily with magical readings.

Like *Astrologaster*, Thomas Tomkis's *Albumazar. A Comedy*, first performed and published in 1615, attacked astrology and its false practitioners. In the play's tale of love and marriage, the astrologer Albumazar, who claims to practise alchemy and necromancy as well, is a fraud who bamboozles his intended victims with verbose gibberish, saying at one point that,

... with scioferical instrument,
By way of *Azimuth* and *Almicantarath*
I'le seek some happy pointe in heaven for you.[39]

---

[35] Melton, *Astrologaster*, p. 10.

[36] Melton, *Astrologaster*, p. 9.

[37] Cerasano, 'Philip Henslowe', esp. pp. 146–50.

[38] Capp, 'Booker, John'.

[39] Thomas Tomkis, *Albumazar. A Comedy* [1620], ed. Hugh G. Dick (Berkeley: University of California Press, 1944), p. 89.

In another place, he coins the wonderfully overblown compound 'Necro-puro-geo-hydro-cheiro-coscinomancy'. Albumazar's material aids include an ephemeris and an increasingly fantastic set of instruments, from the plausible, recently invented 'perspicill' (telescope), to an 'Otacousticon' for improved hearing, which turns out to be no more than an artificial pair of ass's ears, and a 'wind instrument' that can be used to speak in ten languages.[40]

Pulling these threads together can perhaps shed a little more light on perceptions, and thus the practices, of the consultant astrologer. As Melton's and Tomkis's descriptions indicate, the role of instruments such as an astrolabe or telescope, as well as other objects, may be as much about credibility and claims of disciplinary grounding as about actual use within the routine practice of consultation. Such objects might be there to assure the client that the practitioner has the right tools for the job – in this case tools for measuring and studying the heavens – and knows how to use them in the execution of their art. In the case of a celestial globe, which provides a map of the whole sky, one can also imagine it taking on a didactic role, helping to explain the principles or specifics of a case. There is an obvious theatricality in this use of props to create an aura of credibility for the client, a theatricality that Melton and Tomkis play up, both suggesting that such props and the lengthy explanations the practitioner has for them are intended to befuddle the unwary client. Nevertheless, it seems reasonable to imagine consultants like Lilly or Forman deploying instruments in a more didactic/symbolic way, although perhaps not the exaggerated cap, silver wand and mathematical glass.

Secondly, deploying such paraphernalia, printed and manuscript texts included, may not in itself ensure that others will agree that the practitioner is legitimate and virtuous. Rather, an observer might interpret such items as the deliberately deceptive trappings of a deceitful practitioner or of a magical (and thus possibly suspect) one – as Melton certainly does in his extended diatribe and Tomkis does by confounding different practices in Albumazar's eclectic mix of systems. In other words, the tools deployed might equally raise suspicions for those who suspect that something is amiss. This is an important point to bear in mind given the close and generally assumed links that existed between astrology, magic and other often suspect arts in the period.[41] Not all associations were happy ones.

It is not surprising, therefore, also to see images of astrologers within satirical contexts that are strikingly similar to those intended to be taken straight. Indeed, the two might be all but indistinguishable. The best satire, after all, trades on plausibility.

---

[40] Tomkis, *Albumazar*, pp. 80–82, 85, 96.
[41] Thomas, *Religion and the Decline*, pp. 631–35.

In the second part of *Hudibras* (1664–78), a mock-epic poem satirizing the different factions in the English Civil War, Samuel Butler introduces an astrologer, Sidrophel, who is recognizably a parody of William Lilly.[42] Sidrophel is not only dishonest, but also a fool, as Butler illustrates through his (mis)use of instruments and poor celestial knowledge. Outside the astrologer's house, his '*Mansion* prudently contriv'd', for example, stands an obelisk,

> On which was written, not in words,
> But *Hieroglyphick* Mute of *Birds*,
> Many rare pithy Saws concerning
> The Worth of *Astrologick* Learning:
> From Top of this there hung a *rope*,
> To which he fastned *Telescope*;
> The *Spectacles* with which the *Stars*
> He reads in smallest *Characters*.[43]

Yet Sidrophel is no true astrologer and mistakes a light hung from a kite for a celestial omen of the world's end. In this respect he is a 'virtuoso', a dabbler in scientific ideas, full of speculation yet oblivious to the real world around him and easy to fool. In the post-Restoration era that witnessed the foundation of the Royal Society and reports of the extraordinary, and to some people's eyes ridiculous, things they saw fit to investigate and discuss, virtuosi became stock figures in satirical plays and poems, the predecessors of today's absent-minded professor.[44] Butler's conceit goes further as Sidrophel and Hudibras argue about astrology, resulting in a fight in which Hudibras overpowers the astrologer and turns out his pockets. The results are equally revealing:

> First, He expounded both his Pockets,
> And found a *Watch*, with *Rings* and *Lockets*,
> Which had been left with him, t'erect
> A *Figure* for, and so detect.
> A *Copper-Plate*, with *Almanacks*
> Engrav'd upon't, with other knacks,
> Of *Booker's*, *Lilly's*, *Sarah Jimmers*,

---

[42] Curry, 'Lilly, William'.

[43] Samuel Butler, *Hudibras. The Second Part* (London: T. R. for John Martyn and James Allestry, 1664), p. 160–61.

[44] Richard Dunn, *The Telescope: A Short History* (London: National Maritime Museum, 2009), pp. 52–54; Roslynn D. Haynes, *From Faust to Strangelove. Representations of the Scientist in Western Literature* (Baltimore: Johns Hopkins University Press, 1994), pp. 35–49.

And *Blank-Schemes* to discover *Nimmers*;
A *Moon-Dial*, with *Napiers* bones,
And severall *Constellation*-stones,
Engrav'd in *Planetary hours*,
That over *Mortals* had strange powers
To make 'em thrive in *Law*, or *Trade*;
And stab, or poyson, to evade;
In *Wit*, or *Wisdom* to improve,
And be victorious in *Love*.[45]

Again, we have a range of tools the astrologer might use or have around him: the figure as the key artefact of the consultation; the plate with almanacs on it (the equivalent of an ephemeris); calculation aids (Napier's bones); and a time-telling device (the moon dial). What Butler also seems to be doing is deliberately confounding astrological and magical imagery to satirical and polemical effect. There are items not just for making predictions, but also for effecting changes in the world. Butler draws astrological and magical practice deliberately close.

William Congreve's *Love for Love* (1695) also knowingly references William Lilly for its virtuoso/astrologer, Foresight. In the play, Sir Sampson Legend believes he is to be married to the niece of the astrologer and at one point declares,

Where is this old Soothsayer? This Uncle of mine elect? a ha, Old *Foresight*, Uncle *Foresight*, wish me Joy Uncle *Foresight*, double Joy, both as Uncle and Astrologer; here's a Conjunction that was not foretold in all your *Ephemeris*. The brightest Star in the blew Firmament – is shot from above, in a Jelly of Love, and so forth; and I'm Lord of the Ascendant.[46]

Shortly afterwards, Foresight's niece Angelica declares that she hopes her uncle will consent to act in the place of her father and give her away at the marriage. 'That he shall', declares Sir Sampson, 'or I'll burn his Globes'.[47] This is an interesting detail – a rare identification of one of the instruments symbolically associated with astrological practice in images of consultant and other astrologers. Again, one must recognize the satirical intent, but also acknowledge the use of what must have been plausible imagery at the time.

---

[45] Butler, *Hudibras*, pp. 209–10.
[46] William Congreve, *Love for Love. A Comedy* (London: Printed for Jacob Tonson, 1695), p. 60.
[47] Congreve, *Love for Love*, p. 60.

Conclusion

In trying to pin down some practical aspects of the astrologer's art in the early modern period, this chapter has drawn on disparate sources: frontispieces and other images; satirical and polemical writings; astrological casebooks and textbooks. A number of suggestions can be drawn from this brief survey. The surviving records of astrological consultants such as Richard Napier and Simon Forman suggest that the key artefacts within the consulting space – the astrologer's study – were paper-based: manuscripts (horary figures and client notes in the casebooks) and books (ephemerides). Time-telling instruments, sundials in particular, played a role in determining the hour of the question/judgement, although they are rarely mentioned because their role was so unproblematic. The use of other instruments, however, was less obviously active during the consultation. Nonetheless, images of Lilly, Coley and others, as well as polemical and satirical descriptions, suggest a symbolic, legitimising and perhaps didactic role for material aids such as globes (which Lilly certainly owned) and observing instruments, although the suspicion that these and other artefacts were the trappings of false or dangerously magical practices might arise.

The relative stability of the iconography of astrologers, often shown accompanied by a fixed set of attributes including globes and dividers suggests, therefore, an image of 'rhetorical practice' rather than an image of working practice as it routinely took place in the astrologer's workspace. The authority of the practitioner, these images suggest, comes from observations of the heavens based on and codified with instruments and texts, then made concrete through the act of inscribing the horary figure and passing judgement. The astrologer, through his interpretation of the heavenly dispositions with the help of these aids, uses his art to bring stability and order to what might otherwise seem a chaotic universe and to help the client negotiate their future path.

Acknowledgments
I would particularly like to thank Lauren Kassell and Rob Ralley of the Casebooks Project (http://www.magicandmedicine.hps.cam.ac.uk/), who directed me to many of the examples cited.

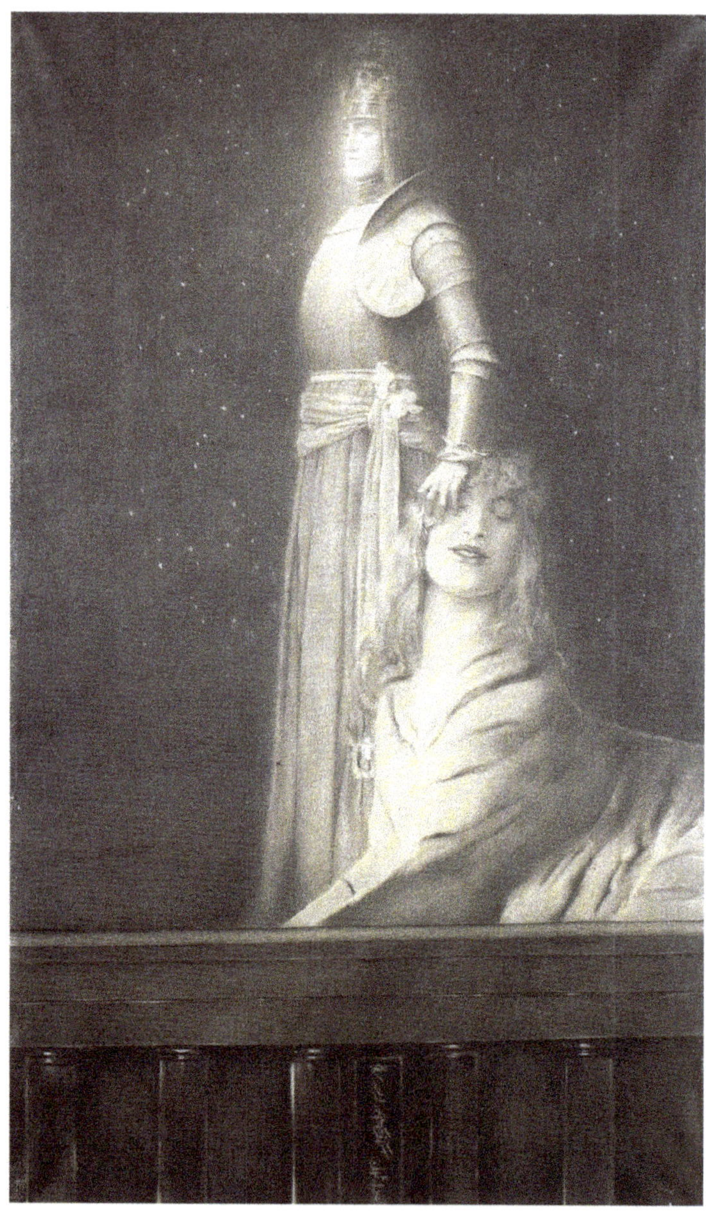

FIGURE 9.1: Fernand Khnopff (1858–1921), '*Avec Verhaeren. Un ange*', 1889, pencil on paper, heigthened with white, 33,1 x 19,8 cm. Previous collection Emile Verhaeren, now in a private collection, Brussels. Published in Michel Draguet, 'Fernand Khnopff 1858–1921'. PhD Dissertation Université Libre de Buxelles: Ghent: Snoek-Ducaju & Zoon; Brussels: Gemeentekredietbank 1995, fig. 280, p. 283.

Twinkling Voices from the Gods: Fernand Khnopff's Use of Stars as Mystical Guides in 'Avec Verhaeren, un ange'

Liesbeth Grotenhuis

ABSTRACT: 'When the lips sleep, the souls awake' Belgian poet Maeterlinck wrote. His words express the quest for a ritual, active silence that was sought throughout the fin-de-siècle. Preferably during night time, the soul of the chosen could learn how to blossom. But how to paint such content? This chapter examines the work 'Avec Verhaeren. Un ange', in which Symbolist painter Fernand Khnopff places a duo against a sparkling sky. The stars relate to the hieroglyph-like signs depicted on the plinth: with the sphinx, it refers to the culture of ancient Egypt. Already the Romans adapted the Egyptian knowledge of astrology, while the early 'scientist' Athanasius Kircher related the wisdom in the zodiac to the even more secret wisdom of the Hellenized Egyptian god, Hermes Trismegistos. Khnopff used this knowledge in a contemporary answer on Péladan's statement: Artist you are priest, you are king, you are magician. His use of the starry sky offered the viewer a key to ancient wisdom.

In 'A total night [...] in dead light', Robert Delevoy describes the setting behind an abstruse constellation of two modest figures (FIGURE 9.1).[1] Due to its prominent role in the unusual composition, the sky demands to be described. Yet the heavens are not as inky dark as suggested: twinkling stars illuminate the scene. Hence my title, 'Twinkling voices from the Gods'.[2] In this paper I shall research the meaning of this subtle light of the starry sky in the drawing by the Symbolist artist Fernand Khnopff (1858–1921).

The nocturnal scene fits the late nineteenth-century context, when personifications from dark till dawn came into fashion; the blue hour afforded artists an unprecedented opportunity to experiment with the monochromatic use of colour to express an Ideal content. The glowing stars not only created special effects, but could also reinforce newly depicted concepts like 'Silence'.[3] As astronomer Camille Flammarion (1842–1925) poetically states in his novel 'Lumen' of 1887: 'What voice is more eloquent than the silence of a star-lit night?'[4] Khnopff used this specific voice

---

[1] 'La nuit est totale [...]. Dans la lumière morte...' Robert Delevoy, 'Rêve Mort et Volupté' as quoted in: Robert Delevoy, Catherine de Croës and Gisèle Ollinger-Zinque, *Fernand Khnopff: catalogue de l'oeuvre* (Brussels: Lebeer-Hossmann, 1987), p. 167.

[2] After: Giordano Bruno, *De Magia*, III, pp.411–12.

[3] Liesbeth Grotenhuis, 'Isis' Fingertips: a Symbolist reading of Lucien Lévy-Dhurmer's "Le Silence", Rosina Neginsky, ed., *The Symbolist Movement: Its Origins and Its Consequences* (Newcastle upon Tyne: Cambridge Scholars Publishing, 2015).

[4] Camille Flammarion, *Lumen* (New York: Dodd, Mead and Co., 1897), p. 169.

to enhance the narrative in the drawing '*Avec Verhaeren. Un ange*'.

The firmament could also be interpreted as possessing another special quality. Vincent van Gogh (1853–1890) compared it to the kind of maps used by sailors throughout the ages. Constellations were employed as a signpost, analogue to the black dots representing cities and villages on the paper map you use when finding your train to Rouen: 'Why would those glowing dots be less accessible to us?' Van Gogh asked.[5] That leads to an even more important question: where would that trip lead? Van Gogh mentioned death as a vehicle, yet there were less radical possibilities for journeying, as Khnopff expressed through his art.

The Rosicrucian Josephin Péladan (1858–1919) and the occultist Eliphas Lévi (1810–1875), specifically provided myths and theories in which the key to an answer can be found for Khnopff's use of the stars. In this context, the theories on astrology and the supposed origins of Egyptian religion are essential, specifically those relating to the goddess Isis and the god/divine human Hermes Trismegistus. To understand Khnopff's stellar rhetoric, three main ingredients must be researched: hieroglyphs, stars, and occult concepts. For the introduction of the Belgian artist, I would like to comment briefly on his rather unconventional photographed portrait (FIGURE 9.2).

This staged portrait provides us with an introduction to Khnopff's multi-layered pictorial language, identical to the one he used in his paintings. Although the artist is the model, he does not occupy a central position, but is relegated to the lower right-hand corner. Another unorthodox feature of the image is that, rather than portraying the artist frontally, it shows Khnopff from the back, his head turned in profile. Behind him is an installation that is identified as being an altar that Khnopff installed in his house in Brussels, a villa built after his own design in 1900.[6]

In the centre, above the letter 'A', hovers a medallion representing the Virgin Mary; the circular shape evokes another layer of reality.[7] On top, the altar contains a single-winged sculpted head of Hypnos, which Khnopff re-used as a model for several of his most important paintings. As the god of sleep and dreams, the anthropomorphic god could assume the guise of a nightingale, which explains his winged

---

[5] Letter 144/123 in: Joachim Pissarro, 'Avond- en nachtthema's in de vroege geschriften van Vincent van Gogh.' ('Evening and night theme's in the early writings of Vincent van Gogh') as quoted in *Van Gogh en de kleuren van de nacht* (Van Gogh and the colours of the night). (Exhibition catalogue Amsterdam: Van Gogh Museum, 2008), p. 59.

[6] Michel Draguet, *Fernand Khnopff 1858–1921* (PhD Diss., Université Libre de Buxelles, 1995), pp. 15, 346–47.

[7] Khnopff used the tondo-shape for several of his paintings, but especially when added as a mirror or a picture within a picture to evokes other worlds; Leslie D. Morrissey, 'Isolation and Imagination: Fernand Khnopff's "I lock my door upon myself"', *Arts Magazine* 53, no. 4 (1978): p. 95.

manifestation. Khnopff not only saw the bust in the British Museum, he 'attached a transcendental meaning to it'.[8] Here we meet Khnopff's vehicle for stellar journeys, the single wing symbolising the representation – or, rather, evocation – of the invisible.

This progressive photograph not only introduces Khnopff as an artist, but affirms his metaphoric approach to telling stories and reveals the meticulous way with which the artist chose to portray himself. In which regard, I would argue that this image is more than just a photographic portrait: Khnopff incorporated himself as part of his own artistry. This changing status was articulated by Sâr Péladan: '*Artiste, tu es prêtre*, artist, you are priest, *tu es roi*, you are king, *tu es mage*, you are magician.' This famous quote is taken from the catalogue of the first exhibition he organized, the *Salon de la Rose & Croix* of 1892.

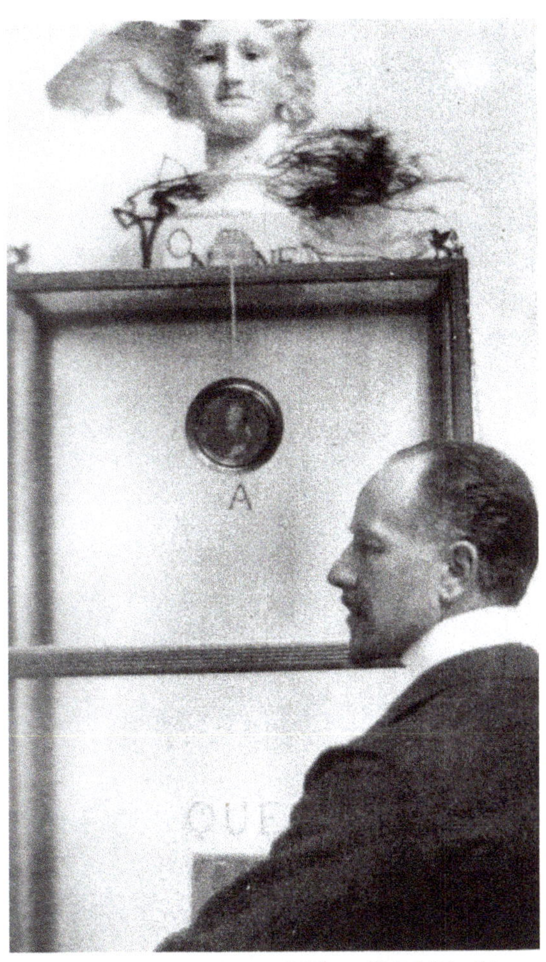

FIGURE 9.2: Portrait Fernand Khnopff. Published in Draguet, *Fernand Khnopff*, fig. 336, p. 347.

The occult frontman was also fond of striking mystical poses; dressed as a high priest, Péladan presented himself often as a messianic figure in diverse painted portraits. In an 1895 canvas by Jean Delville (1867–1953) the golden scroll Péladan holds contains ancient wisdom; its shape refers to Egyptian papyrus-scrolls. Indeed, Pharaonic Egypt was considered an essential source of wisdom by various authors over the ages: although none of the Greek authors had any firsthand knowledge of authentic Pharaonic writing, Plato ascribed both the invention of hieroglyphs and astronomy to the Egyptians, while Plotinus gave a Neoplatonic interpretation of

---

[8] Francine Claire Legrand, 'Fernand Khnopff–Perfect Symbolist', *Apollo* 62, no. 85 (1967): p. 284.

Pharaonic monumental writing that contained an essential truth.[9]

Centuries later, Aleister Crowley (1875–1947) explained the Pharaonic source in his introduction of the influential magician Eliphas Lévi: 'There is no single feature in Christianity which has not been taken bodily from the worship of Isis.'[10] The reference Crowley makes, though, goes mainly back to Hellenistic times, not the older Pharaonic culture; from 300 BCE onwards, the original religion underwent reinterpretation. Hellenistic texts such as those by Apuleius crowned Isis as the Mother of the Universe,[11] confirmed by an inscription explaining her as the embodiment of the cosmic order – 'the one who is all'.[12] As Isis underwent Hellenization, her Pharaonic features and attributes altered, transforming into a new Roman iconography standing in contraposto with her hair styled in corkscrew curls (FIGURES 9.3 AND 9.4). Apuleius specifies her gown with many folds held together between her breasts by the so-called Isis knot: 'along the embroidered border and in the very body of the material there gleamed stars here and there, and in their midst a half–moon.'[13] Roman sculptors could sometimes actually execute her 'cloak of the deepest black, resplendent with dark sheen'[14] in black stone.

As her new images spread, her cult also travelled the length and breadth of the Roman Empire. A magical text from a Greek papyrus originating from Egypt, explains an initiation rite; after a ritualistic death, the insider were to open his eyes, open a door and look into the world of the gods. He would see the mystery of the midnight sun, being the sun-god Re himself, greeting him: 'be hailded [...] bright light, King, Re.'[15] Egyptologist Erik Hornung relates the rite to the Pharaonic sun boat traveling the Netherworld, which in Hellenist times was reinterpreted from the original descent into the Netherworld to an ascent through the spheres of the stars.[16]

---

[9] Erik Iversen, *The Myth of Egypt and Its Hieroglyphs in European Tradition* (Princeton: Princeton University Press 1993); Plotinus, *Enneade*, pp. 5, 8.

[10] Among other eastern deities; Aleister Crowley, 'Introduction', *The Equinox* 1, no. 10 (1913): p. viii; translation of Eliphas Lévi, *La clef des grand mystères suivant Hénoch, Abraham, Hermès Trismégiste et Salomon* (Paris: Felix Algan, 1859).

11 'Rerum naturae parens' Apuleis, *Metamorphoseon*, XI, 5,1 as quoted and translated in John Gwyn Griffiths, *The Isis-Book – Metamorphoses* (Leiden: E. J. Brill 1975), pp. 74, 75. For the role of Isis see: R. E. Witt, *Isis in the Ancient World* (Baltimore: The Johns Hopkins University Press, 1997).

[12] '*Te tibi una quae es omnia dea Isis*' found in the Church of S Maria Vetere in Capua as quoted in Jan Assman, *Egyptian Solar Religion in the New Kingdom: Re, Amun and the Crisis of Polytheism* (1995; repr. London/New York: Routledge, 2012), p. 155.

[13] Apuleis, *Metamorphoseon*, XI, 4,1 in Gwyn Griffiths, *The Isis-Book*, pp. 72, 73.

[14] Apuleis, *Metamorphoseon*, XI, 3,13 in Gwyn Griffiths, *The Isis-Book*, pp. 72, 73.

[15] Papyrus PGM XIII 638 as described in Reinhold Merkelbach, *Isis Regina-Zeus Sarapis*, pp. 175–81.

[16] Erik Hornung, *Das esoterische Aegypten* (München: Beck'sche Verlagsbuchhandlung, 1999) trans.

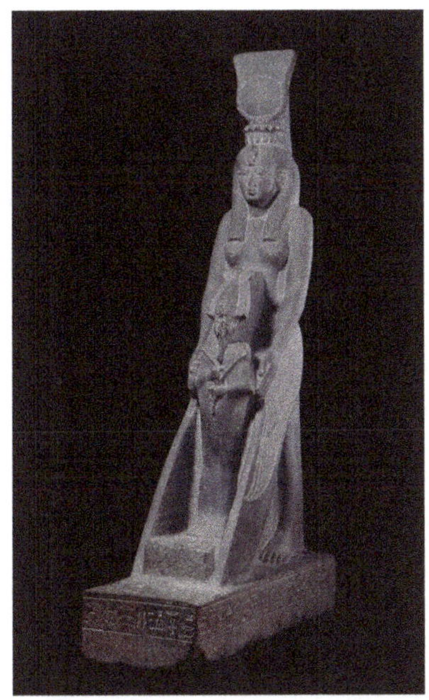 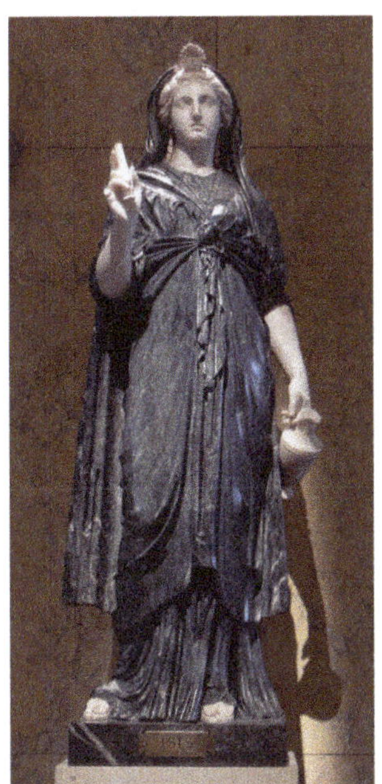

FIGURE 9.3 (left): Pharaonic, 'Statue of Isis protecting Osiris' ca. 590–530 BCE , siltstone, 81,3 x 17 x 47 cm. The British Museum, London. © The Trustees of the British Museum.

FIGURE 9.4 (right): Roman, 'Isis', ca.200–150 BCE, black and white marble, h. 130 cm. Antikensammlung, © Kunsthistorisches Museum Wien,in the public domain, photo: Andreas Praefcke.

Isis evolved further, particularly during the seventeenth century, when Jesuit Athanasius Kircher (1602–1680) granted her a star-studded mantle (FIGURE 9.5). As she roams the earth like a colossus, Isis expresses her ability to move the cosmos, a celestial aspect that the Virgin Mary usurped in her star-halo or a comparable stellar mantle. This illustrates that the gods literally reign above us, the stars as their voices. Kircher presented the Kingdom of Heaven as a hieroglyph itself, the secret language taught by Isis.

While Isis was seen as the teacher, Plato mentioned 'Theuth' as 'some god or some divine man' who not only invented numbers, calculus and the alphabet (hieroglyphs), but also astronomy.[17] 'Theuth' is the Pharaonic god Thoth who appeared in

David Lorton, *The Secret Lore of Egypt: its impact on the West* (Ithaca: Cornell University Press 2001), p. 14.

[17] Plato, *Phaedrus* 274c5–d2; Plato, *Philibus*, 18b–d.

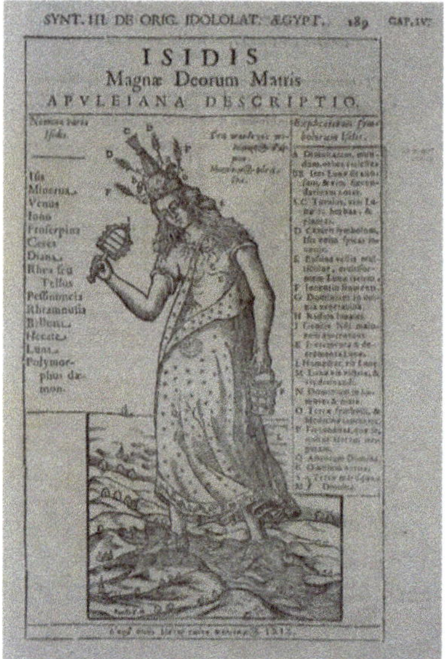

FIGURE 9.5 (above): *Mother Nature, or Syncretized Isis*. From Athanasius Kircher, *Oedipus Aegyptiacus* Rome, 1652.

FIGURE 9.6 (right): Fernand Khnopff, *D'après Joséphin Péladan. Le Vice Suprême*, 1885, pencil, pastel and charcoal on paper, 23 x 12.2 cm. Private collection. Published in Draguet, *Fernand Khnopff*, fig. 269, p. 267.

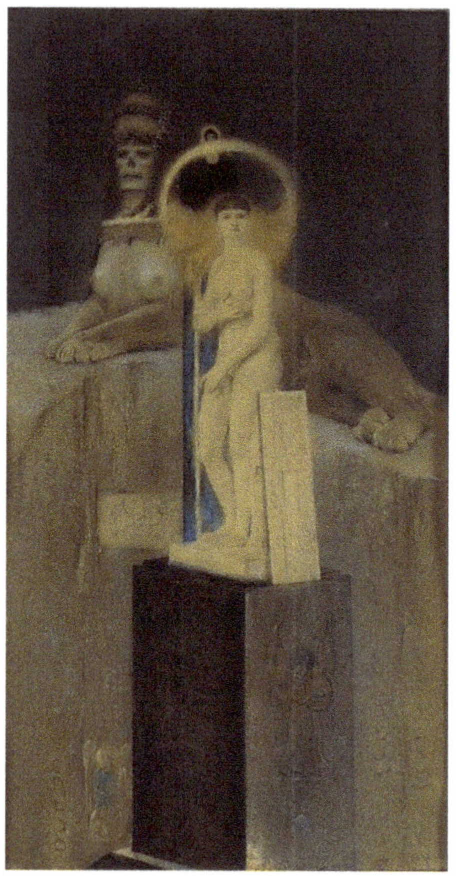

three forms: man, ibis and baboon. He developed into 'thrice-greatest' Hermes Trismegistus, the founder and prophet of hermeticism in which man was conceived as a microcosm.[18] This 'graphic alchemy'[19] established a strong line in esoteric tradition: Charles Dupuis (1742–1809) recognized stars as the source of all world religions, and related the zodiac to the myth of Isis and Osiris.[20] The late nineteenth century astro-

---

[18] Hornung, The secret lore of Egypt, 5; Jacob Slavenburg, 'Hermes in de oudheid.' (Hermes in ancient times), *De Hermetische code* (The Hermetic Code) (Utrecht: Kok ten Have, 2006).

[19] Termed by the French Egyptologist Serge Sauneron.

[20] Charles F. Dupuis, *Origine de tous les cultes ou religion universelle* (Paris: Agasse, 1794), p. 99; Robert P. Welsh, 'Sacred Geometry: French Symbolism and early abstraction.' *The Spiritual in Art: Astract Painting 1890–1985* (Exhibition catalogue Los Angeles: County Museum of Art; Chicago: Museum of Contemporary Art; The Hague: Haags gemeentemuseum, 1987), p. 64.

logical revival was triggered by writings like Ely Star's '*Les Mystères de l'horoscope*' (1887). With its onomastic cabalistic approach, astrology also influenced Péladan, as he emphasizes the references in several of his writings.[21] He especially used the doctrine of 'signatures': celestial phenomena 'as above', were paired to the life on earth, 'so below'. In his best-selling debut novel *Le Vice suprême* of 1884, Péladan expressed his variation of the salvation of man through occult magic of the ancient East. He reacted to the traditional church by 'being modern': having insight in the complete history based on a solid Latin and catholic education and being open for the mystic, involving the higher science of Hermeticism. Its success not only popularised mysticism, but also the artists, who Péladan involved.

It was Fernand Khnopff who was invited to draw the frontispiece for 'The Supreme Vice' (FIGURE 9.6).[22] Here the institute of the church – the papacy – was represented by a sphinx 'powerful in its muscles of the hinter quarter, lust still visible in the breast but the head is exhausted by boredom and tyranny', as described by contemporary critic Emile Verhaeren (1855–1916).[23] The statue of Mary is hardly recognizable behind the radiant halo of her prominent pagan counterpart, the nude Venus. The plinth suggests that she is a sculpture, symbolizing both unobtainable sensuality while simultaneously representing history, as interpreted by Péladan. The rock contains inscriptions, which, although not traceable to any known script, Verhaeren refers to as 'cabbalistic'.[24] The composition is shrouded in mystery, due to the impossibility of deciphering the inscriptions. Similarly enigmatic is the unusual death's head with which Khnopff equipped the sphinx.

There was, however, another Belgian artist who actually illustrated Péladan's novel the year before: Felicien Rops (1833–1898) had been Péladan's regular illustrator until now.[25] Khnopff was a great admirer of his work. When seeing Rops' oeuvre, we immediately recognise the dark elements as Khnopff's source for his depiction of the sphinx head as a skull, and the regular depicted antique reliefs used as a plinth.

---

[21] Christopher McIntosh, *Eliphas Lévi and the French Occult Revival* (Albany: SUNY Press, 2011), p. 167.

[22] Fernand Khnopff, 'After Joséphin Péladan' or 'The Supreme Vice' 1885, pencil, pastel and charcoal on paper, 23 x 12,2 cm. Private collection.

[23] '*Oh! la mortuaire image de papauté sur un corps moitié lion, moitié sphinge. Puissance encore dans les griffes et les muscles et la croupe, volupté encore dans la gorge, mais la tête, tellement ennuyée, creuse, séculaire, immense d'usure et de tyrannie!*' Emile Verhaeren, 'Silhouttes d'artistes: Fernand Khnopff', *L'Art Moderne* (1886), p. 41; as quoted in: *Fernand Khnopff et ses rapports avec la Secession viennoise* (Exhibition catalogue Brussels: Musées Royeaux des Beaux-Arts de Belgique, 1987), p. 31.

[24] '*un socle où se lisent des caractères cabalistiques*' Verhaeren, 'Silhouttes d'artistes' as quoted in: *Fernand Khnopff et ses rapports avec la Secession viennoise*, p. 31.

[25] Felicien Rops, 'The Supreme Vice' 1884, Etching, 15,9 × 26,8 cm.

Rops used this characteristic layout in *Pornocratès* with a figure on an inscribed architectural structure seen from a low point of view (FIGURE 9.7). In terms of composition, the outline of an upraised female with a beast against a starry sky, bears such a close relationship to Khnopff's later work '*Avec Verhaeren*', that I would argue that Khnopff was responding directly to this drawing. Yet there are also essential differences: Rops's frivolous feathered hat transformed into stern armour. The sensuous flowers embellishing the hair have become a lily, symbol of virginity, now attached to the waistband of Khnopff's knight. In that respect, the contrast could not be greater: Khnopff's knight-like figure with the sphinx is an enlightened answer to the blindfolded nude that walks the hog. Khnopff now mystified the pornographic content drenched in provocative Satanism, Rops' signature style. When Péladan excluded Rops from his retinue for having created an all-too blasphemous illustration, Rops developed a life-long hatred of Khnopff.

In its capacity as an enigmatic response, the work asks why Khnopff decided on a sphinx as the pig's counterpart. She is interpreted as sensuous with a body 'run to seed',[26] yet this sphinx is too complex a creature to justify this one-dimensional reading. She is not a hog. As a hybrid creature, she is inherently a blend of man and beast – the body of an animal and the brains of a human, as aptly illustrated by the intellectual battle of the riddle between the sphinx and Oedipus. Her pendant, the antique Egyptian sphinx, is traditionally used in western art as the keeper of ancient wisdom.

Esoteric groups indeed used the sphinx in this context; an eighteenth-century Masonic medal by Martin Folke presents a reclining Egyptian sphinx before two Roman monuments: the steep pyramid of Gaius Cestius and the Aurelian wall (FIGURE 9.8). A lunar crescent is carved on her shoulder, relating her to the cosmos which identifies her as a creature of Isis.[27] From these occult connotations, Khnopff developed the elusive qualities of this sphinx to a deeper level. It was Péladan again who provided new content for Khnopff's concepts: in '*La Jeune Belgique*' (1886) (the young Belgian), the superhuman genius of the artist was presented as a 'star child', in contrast to ordinary people living on earth. This duality forms the starting point for Khnopff's pencil drawing 'Paganism' (ca. 1910): the bottom part of the picture plane shows the lower god Pan backed up by a lustful crawling nymph, topped by an androgynous person with closed, silent eyes holding a laurel branch.[28] This figure is

---

[26] Draguet, *Fernand Khnopff*, p. 284.

[27] David Ovason, *The Secret Zodiacs of Washington DC: Was the City of Stars Planned by Masons?* (London: Arrow Books 2000), p. 284.

[28] Fernand Khnopff, 'Paganism', ca. 1910, coloured pencils on paper, 36 x 30 cm.

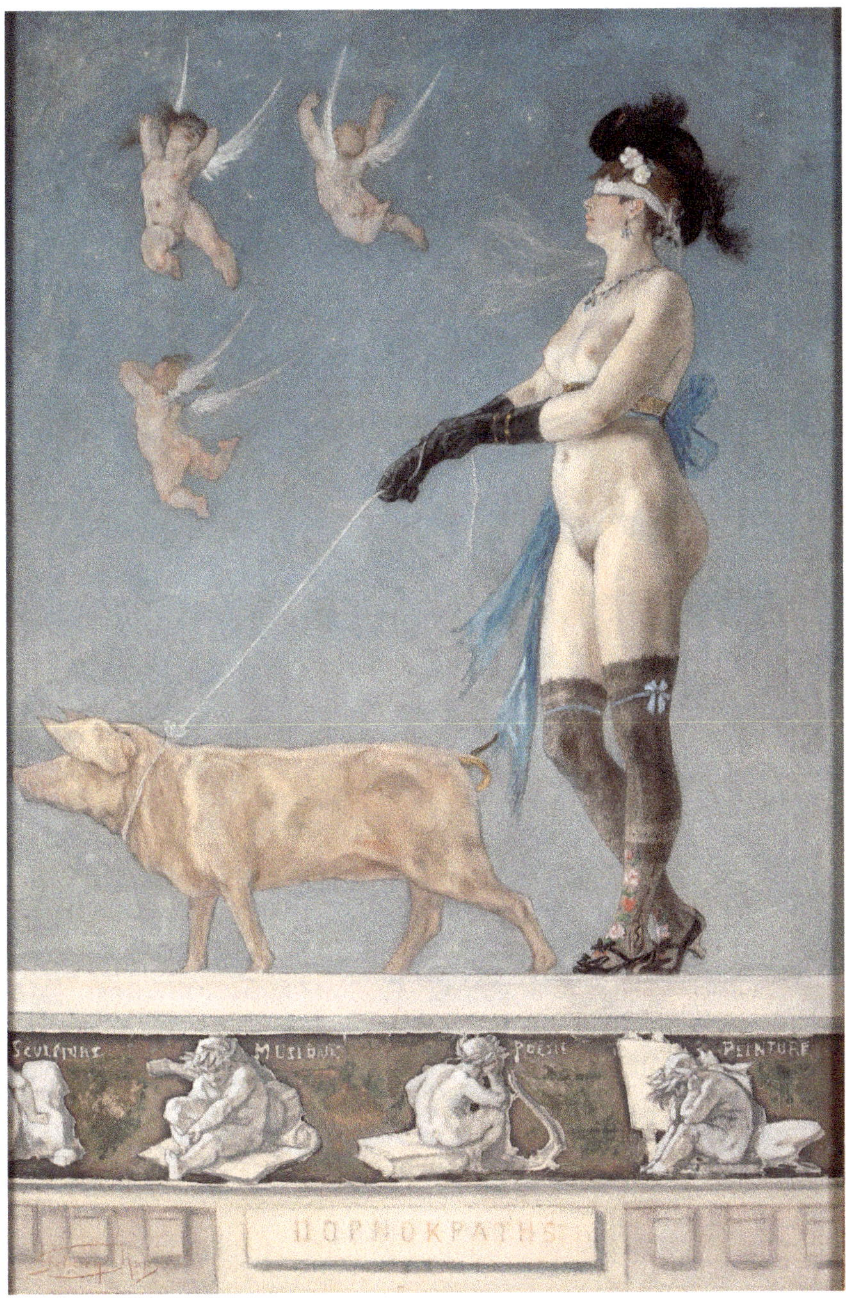

Figure 9.7: Félicien Rops, '*Pornocratès* or *La Dame au cochon*', 1896, coloured engraving by Albert Bertrand after an aquarel by Rops, 69 x 45 cm. Coll. musée Félicien Rops, Province de Namur, inv. G E853 © musée Rops Namur.

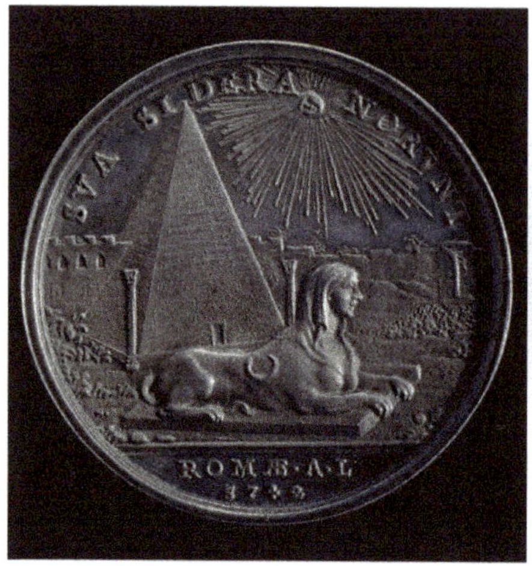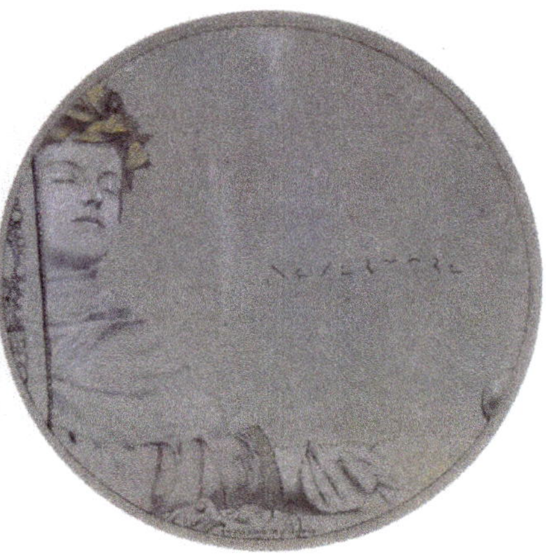

Figure 9.8: Masonic, '*Sva sidera norvnt*', 1742, bronze, d. 3,7 cm. British Museum, London. © The Trustees of the British Museum.

Figure 9.9: Fernand Khnopff, '*Une rêveuse*' (A dreamer), ca. 1900, chalk on paper, d. 13,4 cm. Private collection. Published in Draguet, *Fernand Khnopff*, fig.3, p. 9.

indicated as 'immortal paganism', referring to the truthful essence of religion.

The closed eyes and the laurel are also evident in the drawing 'The Dreamer', in which the laurel is now worn as a wreath. Khnopff placed the figure to the extreme left-hand and lower edges of the circular canvas, leaving an empty centre where broken letters can be deciphered as 'never more' (FIGURE 9.9). The riddle-like qualities are extended to the pose of the figure who is called '*Une Rêveuse*', a dreamer, and thus recalling Hypnos, as well as the protagonists with their closed eyes in '*Avec Verhaeren, Un Ange*'. The state of dreaming is related to the night and the stars. Verhaeren articulates its essence, asserting that ideas were particularly formed during the nocturnal hours because, in the morning, 'the tender twinkle of the star became invisible'.[29] In his article of 1886, Verhaeren described Fernand Khnopff as a close-mouthed individual, suspecting such inner, dusky discussions.[30] In other words, night-time revealed stars, specifically the subtle light, which Verhaeren recognized as the moment of inspiration. This idea could have pleased Khnopff who was, after all, alluded to by Péladan as a star child. But above all Khnopff was a registered be-

---

[29] '*oui, la nuit quand l'idée s'ébauche et qu'il faut la saisir et, sautant du lit, la clouer sur le papier ou sur la toile. Le lendemain, ce frêle tremblement d'étoile, on ne le verrait plus*.' Verhaeren, 'Silhouttes d'artistes' as quoted in: *Fernand Khnopff et ses rapports avec la Secession viennoise.*

[30] Emile Verhaeren, 'Silhouettes d'artistes. *Fernand Khnopff*' *L'Art moderne* (12 septembre 1886).

liever in astrology.[31] One of the ceilings of the house he built for himself, was decorated with the constellation of Libra, the sign under which Khnopff was born. Neither the ceiling, nor a picture of it, have survived.

The first part of the title '*Avec Verhaeren. Un Ange*' refers to the name repeatedly quoted to explain different aspects of Khnopff's background. Indeed, the poet Emile Verhaeren launched the careers of several young Symbolist painters by writing biographies and articles in magazines like '*La Jeune Belgique*' and '*L'Art Moderne*'. By paying attention so early in his career, Khnopff rewarded Verhaeren with this pencil drawing named after him in 1889; now in a private collection, measuring 33,1 x 19,8 cm. Khnopff must have been very pleased with the small drawing, for he copied the composition in pastel the same year, and enlarged it drastically to 139 x 79 cm. After having designed a silver frame for it, he kept the piece in his own collection. It also functioned as a prop in studio photos in which Khnopff, dressed in a formal dress-suit, acts as if adding the finishing touches to the work.

The theme also appeared in a third variation, this time in the form of a photograph (by Arsene Alexandre, 29,5 x 18,4 cm. Royal Library Albert I, Brussels) that was hand-coloured by Khnopff, also in 1889. When comparing the works, an analysis of the differences prompted me to decipher the originals in more detail. In the photo, Khnopff covered the entire sky with clusters of stars. But in both drawings, the majority of the sky surrounding the duo is sprinkled with stars. This leaves a pitch-black strip to the left of the picture plane; in addition to the missing white stars, the black is even blacker, as if the stars illuminate the sky more subtly than the glowing face of the standing knight.

The shady side of the firmament led me to a detail in the painting 'The Adoration of the Magi' of ca. 1495 by Raffaello Botticini, (1477–ca.1520) (FIGURE 9.10). The ruin, in which the scene is set, reveals an interesting configuration in the apse: a minutely detailed pillar on which a broken statue represents the old, pagan world in collapse. Yet this pre-Christian wisdom never completely vanished: the plinth's central panel shows the adoration of Hermes, depicted with the head of the baboon, one of the manifestations of the original god Thoth, which evolved into Hermes Trismegistus.[32] A star is added to confirm his qualities as the designer of the hieroglyphs, analogue to the constellations of the zodiac.

But I was also struck by Botticini's use of shadow, not only the shadow cast by the pillar, which now seems bereft of its statue. The cosmic canopy lacks golden stars

---

[31] Jeffery W. Howe, *The Symbolist Art of Fernand Khnopff* (Ann Arbor: UMI Research Press 1982), p. 148.

[32] Thanks to Liana De Girolami Cheney for discussing this detail.

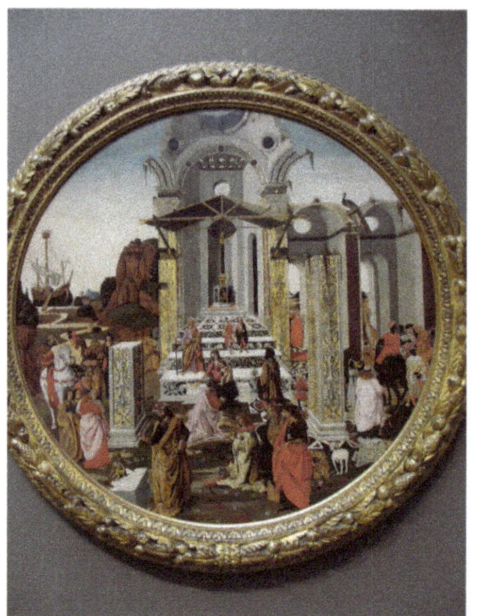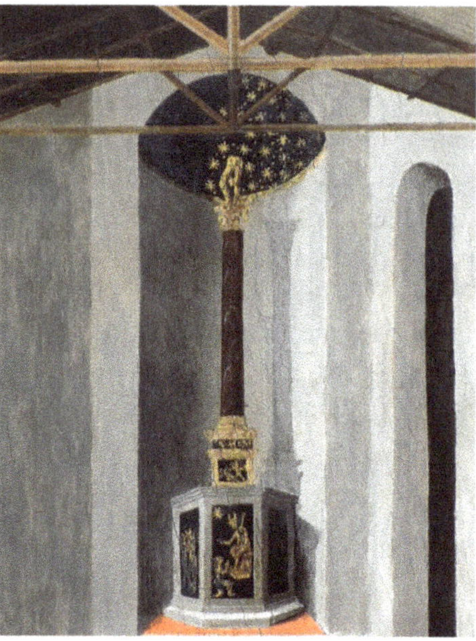

FIGURE 9.10 (left): Raffaello Botticcini (Attributed to) (1477–ca.1520), 'The Adoration of the Magi', ca. 1495, tempera on panel, d. 104,2 cm. Mr. and Mrs. Martin A. Ryerson Collection, 1937.997, The Art Institute of Chicago. FIGURE 9.10.A (right) Detail.

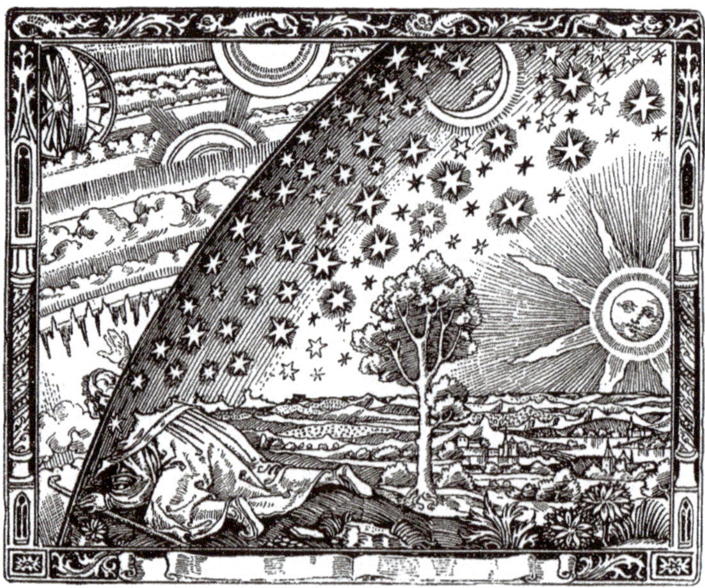

FIGURE 9.11: Camille Flammarion (1842–1925), 'Shepherd peering into the void at St Catherine's wheel' woodcut, *L'Atmosphère: Météorologie Populaire* (Paris: Hachette, 1888), p. 163.

on the left side. In the Botticini, this seems logical, creating the effect of a hollow architectural dome, raising the question whether this lack of light also affects the heavens, too. In any case, shadows contain multiple meanings: art historian Ernst Gombrich offers the following: shadows are not part of the real world.[33] It is exactly this aspect that offers a unique possibility to represent the non-visible, a subject Symbolist painters were eager to depict. By its form and placement in relation to the two figures in 'Avec Verhaeren. Un Ange' the deep black strip seems to suggest that it might represent a doorway – an entrance to another world.

This idea was visually triggered by an engraving as published in Flammarion's 'L'atmosphere: meterologie Populaire' of 1888 in which a 'Medieval missionary' found the edge 'where the sky and earth touch' and 'passes under the roof of heaven' (FIGURE 9.11). Moreover, in the print, the male figure's head and shoulders penetrate the celestial dome, and he peers into the Ideal beyond the earthly world. This essentially different world is depicted in almost abstract circular forms, partly radiating between diverse layers that suggest a broader space than either heaven or earth. Where this man is clearly experiencing a visible vision, Khnopff's knight does not even open her eyes. How should these wide shut eyes be interpreted? The function of her companion, the sphinx, provides a pointer.

In view of the sphinx's traditional role as guard, the dark strip could easily be interpreted as a passage. Khnopff did not depict her as the well-known prototype, though, being an architectural element based on antique stone sculpture. Nor does he refer to the impassive reclining Pharaonic sphinxes with the characteristic royal headdress, nor to two heraldic Assyrian bull sphinxes Péladan used as a cover for the catalogue introducing his first Rosicrucian Salon. Simply acting the part of defender would not have been clear enough for Khnopff's Symbolist message. In using the sphinx (in contrast with Rops's hog), Khnopff evokes deeper layers. In addition to being a hybrid creature with mythical references, the sphinx also offered a new meaning.

In magician Lévi's encyclopaedic overview of 1861, his sphinx was composed as a sitting winged lion with female breasts and head, wearing the Pharaonic nemes, thus fusing the Greek prototype with the Pharaonic origins (FIGURE 9.12). The lemma refers to the riddle the Theban sphinx asked: 'What animal is it that in the morning goes on four feet, at noon on two feet and in the evening on three feet?' Oedipus answers correctly: 'man himself: in childhood he crawls upon hands and knees, in

---

[33] E. H. Gombrich, *Shadows: The Depiction of Cast Shadows in Western Art*. Exhibition catalogue (London: The National Gallery, 1995), p. 17.

manhood he stands erect and in old age he supports himself with a cane'.[34]

Lévi recognizes 'still another answer to the riddle of the sphinx', using a kabbalist interpretation as the key to past, present and future.[35] The first part of the riddle refers to number 4 – the creeping – and represents ignorant man and infant humanity, the 2 to intellectual man and evolving humanity, followed by 3 as the spiritual man or illuminated magus with the staff of wisdom; 'The sphinx is therefore the mystery of Nature, the embodiment of the secret doctrine and all who cannot solve her riddle perish.' The pseudo-hieroglyphs on Lévi's pedestal corroborate the essence of her knowledge, analogue to Khnopff's letter signs. But where the plinth defines Lévi's sphinx as a sculpture, Khnopff animated his creature, thus personalizing her. How to relate this content to my hypothetical doorway? To let the star twinkle, Khnopff reduced his already subdued use of colour even more on the left side, where it is entirely. Here, light is completely absent. This darkness is a serious indicator of where the passage should lead. This contrast is played off in another occult source Lévi advocated as a key to 'understand all nature's hieroglyphs': tarot cards.[36] Antoine Court de Gébelin was the first to relate this 'road to wisdom' to Egyptian theology in *Le Monde Primitif* (1781);[37] the 'Papesse' (later the 'High Priestesses') was identified as Isis. In the Wirth deck (1889) she sits on a sphinx throne between a pink/red and a blue column, the two indestructible pillars of Boaz and Jachin on which prelapsarian Adam or Seth had written all the knowledge which the Egyptians derived and taught to the Greeks.[38] In the later Egyptianesque desk from 1863 the card transformed to 'The gate of the sanctuary' where Isis holds (secret) knowledge in the form of a scroll, partly hidden under her veil (FIGURE 9.13).[39] In this design, one column is spotlighted, the other shaded, literally illustrating the adage 'through darkness to light'. This

---

[34] The phenomenon of the sphinx's riddle was first mentioned by Pindar, Fr.117 d. Apollodorus revealed the content with the three phases of life, as well as the origin 'from the Muses' Apollodorus, Bibliothèkè. III.v.7

[35] Eliphas Lévi, 'Introduction', 1896, as quoted in: Christopher McIntosh, 'Eliphas Lévi.' in: Christopher Partridge, eds., *The Occult World* (Abingdon, New York: Routledge 2014), p. 225.

[36] Lévi, 'Introduction', *The Occult World*, p. 225. See also Mark Filipas, 'A History of Egyptian Tarot Decks', *Fourhares Tarot Studies Newsletter* August 2008 at http://newsletter.tarotstudies.org/2008/08/history-of-egyptian-decks/ [Accessed 24 September 2018].

[37] In volume VIII; Mike Sosteric, 'A Sociology of Tarot', *Canadian Journal of Sociology* 39, no. 3 (2014).

[38] E. H. Gombrich, 'Icones Symbolicae', *Gombrich on Renaissance: Symbolic Images* (London: Phaidon, 1972), pp. 149–50.

[39] Both the Paul Christian deck with designs by Edmond Billaudot deck from 1863 and the deck inspired by R. Falconnier with designs by Otto Wegener from 1896 contain 22 cards that were supposed to be cut out the book forming a deck: P. Christian, *L'homme rouge des tuileries: de 22 figures kabbalistiques* (1863; repr. Paris: Guy Tredaniel 1977).

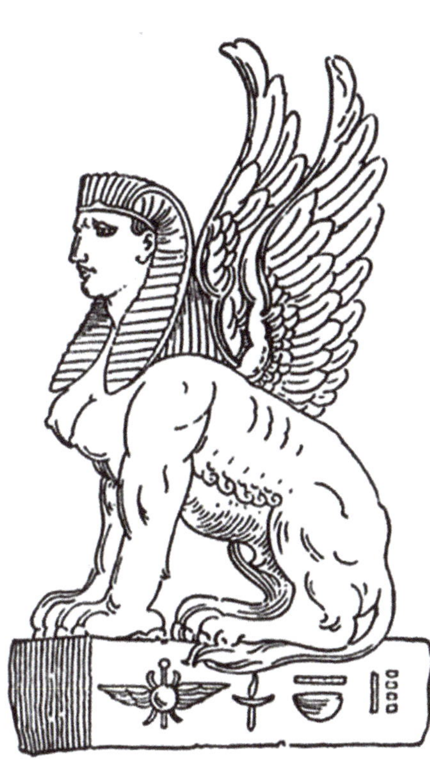

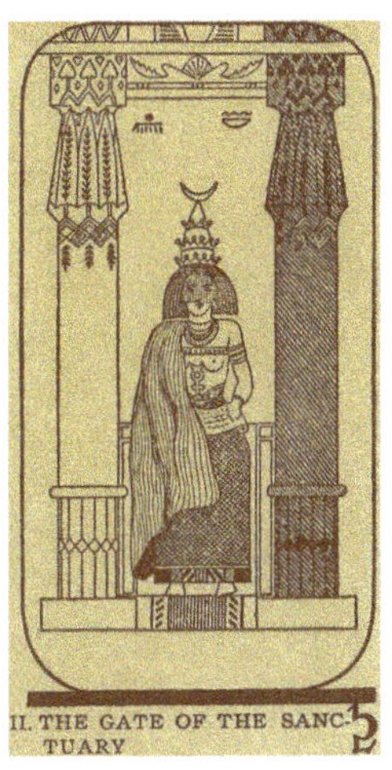

Figure 9.12 (left): Éliphas Lévi (1810–1875), 'Oedipus and the Sphinx'. *Les Mysteres de la Kabbale ou L'Harmonie occulte des Deux Testaments* (1861; repr. Paris: Guy Trédaniel 1991), p. 42.

Figure 9.13 (above): Maurice Otto Wegener (1849–1924), 'The gate of the sanctuary' from R. Falconnier, *Les XXII lames hermétiques du tarot divinatoire* (1896; repr. Nice: Collection Belisane 1976).

depiction illustrates the spiritual journey Khnopff's depicted knight advocates.

The absence of light not only paralleled a primeval state of ignorance; crossing this state as a ritual death was part of initiation, leading to the ultimate wisdom of the initiated. Or, in the words of Lévi: 'To pass the sphinx is to attain personal immortality.' Now who was the one to enter? Khnopff was very specific, as reads the motto on his personal ex libris: '*On n'a que soi*', one only has oneself. The spiritualization of mankind, as explained by Lévi, was an individual quest.

It is precisely this maxim that is acted out in the portrait photograph (FIGURE 9.2): Hypnos's plinth is described with the first part '*on ne*', the 'a' is visible under Mary in the small tondo, finished with the '*que*' besides the arm of the painter, who now serves as a pictogram for the '*soi*-part' of the maxim. Now the Catholic medallion can be understood in the broader context of the origin of religions, after all, 'every religion is born in spirituality. The principle of paganism should be sought in the human heart,' continues Baron Portal in his writing of 1837, an essence that can be revealed in the world of dreams, as Hypnos shows in this assemblage.[40] It can become active when initiated.

When looking again at the constellation formed by the knight and the sphinx in 'Avec Verhaeren. Un Ange', I propose to identify the duo as 'one self' too, as two opposing yet paired characters that offers a solution when fused harmoniously. As Verhaeren explains: 'We have to use sphinxes, ancient kings and fairy queens, legends and epopees to make ourselves audible'.[41] Here the two opposites are not separated by a frame like in 'Paganism'; they are intermingled as two aspects that had to become one in order to enter the gateway of higher knowledge.

Even the figures themselves unite contradicting aspects: as the title indicates, the standing female is an angel, yet she appears in a suit of armour and without wings. As a standing, slightly radiating star child she contrasts with the lying sphinx on the earth. Her 'sensuous smile' indicates the pre-eminently dangerous element of the female: she seduces men. As Péladan argued: 'man is possessed by woman, woman is possessed by the devil';[42] articulating the late nineteenth century male consideration of women being substantially different creatures. Yet in his spiritual development, man needed to pass this serious obstacle; in 'Avec Verhaeren. Un Ange' the distraction is indicated by the sensuous smile. It is remarkable how Khnopff indicated the female aspect: instead of presenting a one-dimensional temptress of our world, he depicts a complex mythical creature. A creature that evolved over four years, compared to his sphinx of 'The Supreme Vice'. The skull is gone, as is the female breast, 'in which lust was visible': Khnopff's new sphinx seduced in a more modest way, leaving room to combine her attraction with her primeval wisdom more trustworthy.

Again, an ex-libris elucidates Khnopff's mentality: under a female's head (her eyes closed) is written 'mihi', meaning 'myself'. Her features are easily identified with the model Khnopff used over and over again, Marguerite Khnopff. By presenting his sister as 'myself', Khnopff fuses his own male personage with that of a female with similar facial features into one, thus overcoming the human failing, being either male or female. Art had the capacity to depict the Ideal state of androgyny. Indeed the answer may be found in Péladan's notion of the hermaphrodite in what McIn-

---

[40] Frédérick Portal, *Des Couleurs Symboliques*, 1837; trans. W. S. Inman, *Symbolic Colours in antiquity, the middle ages and modern times* (London: George Woodfall, 1845), p. 4.

[41] Verhaeren, 'Silhouttes d'artistes', in *Fernand Khnopff et ses rapports avec la Secession viennoise*, p. 22; '… ce seront les sphinx, les anciens rois et les reines fabuleuses, et les légendes et les épopées qui nous serviront à nous faire comprendre'.

[42] Josépin Péladan, 'Les Maîtres contemporaines: Félicien Rops', in *La Jeune Belgique* 15 (1885), as quoted in: Robert Pincus-Witten, 'Occult Symbolism in France: Joséph Péladan and the Salons de Rose + Croix', (PhD Diss., University of Chicago, 1968), p. 58.

tosh describes as the fusion of 'voluptuousness and intelligence, of the active and the completative faculties'.[43] McIntosh argues that for Péladan entities that contained double bodies or personalities were of special occult significance.

In conclusion, Khnopff's drawing '*Avec Verhaeren. Un Ange*' presents two ambiguous figures; a wingless angel wearing the outfit of a knight paired with a seductive sphinx that simultaneously represents wisdom. The constellation shows different aspects of individual spiritual development; all polarities should be harmonized, and to contemplate the essence of existence, religious origins and the state of dreaming must be fathomed. Both pseudo-hieroglyphs and the starry sky hold the promise of secret knowledge accessed by walking the path of initiation, to be started with the entering of absolute darkness and illustrating Khnopff's continuous search for a contemporary take on a traditional symbol.[44]

---

[43] McIntosh, *Eliphas Lévi and the French Occult Revival*, p. 165.

[44] Ludwig Hevesi, *Fernand Khnopff, the Master Mystic from Brussels*. Republished in: *Fernand Khnopff et ses rapports avec la Secession viennoise Exhibition catalogue* (Brussels: Musées Royeaux des Beaux-Arts de Belgique 1987), p. 159.

THE SOPHIA CENTRE

The Sophia Centre for the Study of Cosmology in Culture is a teaching and research centre within the Faculty of Humanities and the Performing Arts at the University of Wales Trinity Saint David. The Centre works from a humanities and social science perspective and encompasses research styles and methodologies from anthropology, history, philosophy, religious studies and sociology. Its academic goals are:

- 'to pursue research, scholarship and teaching in the relationship between astrological, astronomical and cosmological beliefs and theories, and society, politics, religion and the arts, past and present' and

- 'to undertake the academic and critical examination of astrology and its practice'.

The Centre's wider goal is stated in its title – to 'study cosmology in culture'. This enables us to tackle a wide range of topics, from Egyptian sky religion and Babylonian astrology, to astronomy in surrealist painting, astrology in contemporary culture, UFO abduction and the politics of the space race. The Centre promotes research in the subject area, holds seminars and conferences, is associated with Sophia Centre Press and the publication *Culture and Cosmos*, and supervises PhD students.

To find more about our work please visit http://www.uwtsd.ac.uk/sophia/

Dr Spike Bucklow is a Senior Research Scientist and Teacher of Theory at the Hamilton Kerr Institute, Fitzwilliam Museum, University of Cambridge. Bucklow trained as a chemist and worked in the film industry before moving into the conservation of Old Master paintings. He holds a PhD in art history and now teaches at the Hamilton Kerr Institute, University of Cambridge. His publications include *The Alchemy of Paint* (2009), *The Riddle of the Image* (2014) and *Red: The Art and Science of a Colour* (2016).

Nicholas Campion is a historian and anthropologist, specialising in the cultural contexts and consequences of cosmology, astronomy and astrology. He is Associate Professor in Cosmology and Culture at the University of Wales Trinity Saint David, where he is also programme director for the MAs in Cultural Astronomy and Astrology, and Ecology and Spirituality His books include the two-volume *History of Western Astrology* (London: Continuum 2008/9) and *Astrology and Cosmology in the World's Religions* (New York: New York University Press, 2012) and he is the editor of *Culture and Cosmos*, the journal on the history of cultural astronomy and astrology. He submitted various articles to the *Biographical Dictionary of Astronomers*. His recent papers include 'Cosmos and Cosmology', in Robert Segal and Kocku von Stukrad (eds), *Vocabulary for the Study of Religion* (Leiden: Brill, 2015) and 'The Importance of Cosmology in Culture: Contexts and Consequences', in Jesse Abraao Capistrano de Souza (ed.), *Cosmology* (Rijeka: InTech Open, 2017). His current projects include the six volume *Cultural History of the Universe* (Bloomsbury, forthcoming), for which he is General Editor (with Richard Dunn).

Ruth Clydesdale is an author, astrologer and independent researcher, holding an MA in Renaissance Studies from Birkbeck College, University of London, and the Diploma of the Faculty of Astrological Studies. She is the author of *Hermes: The Soul's Companion* (2009), *Secret Wisdom* (2009) and *The Happiness Handbook* (2011), and her articles on astrology have been published in the UK, US, and Europe.

Richard Dunn is Senior Curator and Head of Science and Technology at Royal Museums Greenwich. He is a historian of science by training, having completed a PhD on astrology in Elizabethan England. He has worked in museums for many years and has published on the histories of navigation, scientific instruments and astrology. His recent publications include *The Telescope: A Short History* (2009) and *Finding Longitude* (with Rebekah Higgitt, 2014).

Dr Martin Gansten is a Sanskritist and historian of religion specializing in astrological and divinatory traditions as well as in Indic religions. He received his doctorate from Lund University, Sweden, where he is now *docent*. He has published on the history of astrology in several cultures and epochs, from Hellenistic Egypt to modern Europe, in addition to various aspects of Hinduism. His current research deals with the medieval and early modern Indian reception of Perso-Arabic astrology and related knowledge systems.

Liesbeth Grotenhuis is a Dutch art historian, earning her MA art history (Modern Art and Egyptian Art) from the University of Groningen and Leyden. Specialising in nineteenth-century art and particularly the Egyptian influences on it, her current research is on the iconographic motif of the Sphinx in Academism and Symbolism. Diverse related papers were presented on different conferences and articles are published and in press.

John Meeks holds an MA in German Literature from the University of Western Ontario. After teaching for eleven years at Emerson College in England, he worked as an upper school teacher in Switzerland. He is now a freelance lecturer and writer. His areas of special interest and research include astronomy, mythology and the history of consciousness as expressed in music, the visual arts and literature.

Suzanne Nolan is a PhD candidate in completion in the School of Philosophy and Art History at the University of Essex. She completed her BA, and MA in Pre-Columbian Art and Architecture at Essex. Nolan's research focuses on the Classic Maya civilisation (250–900 CE), and its political organisation. She specialises in late-Classic political transformations at Yaxchilán, a ceremonial centre situated in Chiapas, Mexico. She teaches workshops on Maya hieroglyphic writing, and lectures on pre-Columbian culture and contact period history. Suzanne has presented at international conferences, including the ASA in new Delhi and EMC in Brussels.

Dr Claudia Rousseau, born and raised in New York City, completed her BA at Hunter College (cuny), and her MA and PhD in Art History at Columbia University in New York. Dr. Rousseau is Professor of Art History at the School of Art and Design at Montgomery College where she has been teaching Renaissance and Modern Art since fall 2001. Rousseau has written extensively about astrological imagery in Italian Renaissance art, beginning with her doctoral dissertation *Cosimo I de' Medici and Astrology: The Symbolism of Prophecy.*

Micah Ross is a post-doctoral researcher under Kuang-tai Hsu at the National Tsing Hua University in Hsin Chu, Taiwan. He has studied Sanskrit astronomy and astrology at Kyōto Sangyō University and served as a researcher at the Institute d'Etudes Avancées and within the REHSEIS workgroup of Université Paris 7, both in Paris. He graduated from the Department of the History of Mathematics at Brown University in Providence.

Dr Jennifer Zahrt researches the many forms of astrology emergent across human cultures past and present, with a special focus on early twentieth century German culture. In 2017 she was awarded a five-year appointment as an Honorary Research Fellow at the Sophia Centre for the Study of Cosmology in Culture at the University of Wales Trinity Saint David. She is the founder of Revelore Press, creative director of the Sophia Centre Press and the deputy editor of the peer-reviewed journal *Culture and Cosmos*. She has taught and lectured domestically and internationally in places such as Canada, the United Kingdom, and Germany. She is currently based out of Seattle, wa.

Index